Oxford International Primary Maths

Tony Cotton

Caroline Clissold
Linda Glithro
Cherri Moseley
Janet Rees

Language consultants:
John McMahon
Liz McMahon

2

OXFORD
UNIVERSITY PRESS

OXFORD
UNIVERSITY PRESS

Great Clarendon Street, Oxford, OX2 6DP, United Kingdom

Oxford University Press is a department of the University of Oxford. It furthers the University's objective of excellence in research, scholarship, and education by publishing worldwide. Oxford is a registered trade mark of Oxford University Press in the UK and in certain other countries

© Cherri Moseley 2014

The moral rights of the authors have been asserted

First published in 2014

All rights reserved. No part of this publication may be reproduced, stored in a retrieval system, or transmitted, in any form or by any means, without the prior permission in writing of Oxford University Press, or as expressly permitted by law, by licence or under terms agreed with the appropriate reprographics rights organization. Enquiries concerning reproduction outside the scope of the above should be sent to the Rights Department, Oxford University Press, at the address above.

You must not circulate this work in any other form and you must impose this same condition on any acquirer

British Library Cataloguing in Publication Data
Data available

978-0-19-839460-0

10 9 8 7 6 5 4

Paper used in the production of this book is a natural, recyclable product made from wood grown in sustainable forests.
The manufacturing process conforms to the environmental regulations of the country of origin.

Printed in Great Britain by Bell and Bain Ltd, Glasgow

Acknowledgements
The publishers would like to thank the following for permissions to use their photographs:

Cover photo: Denis Burdin/Shutterstock, P1: MIXA/Getty Images, P11a: Vitalinka/Shutterstock, P11b: Keith Levit/Shutterstock, P11c: David Steele/Shutterstock, P11d: Alexandra Gl/Fotolia, P11e: Barragan/Shutterstock, P11f: iStock.com, P11g: Nagib/Shutterstock, P123: Masterfile, P159a: Matteo Festi/Shutterstock, P159b: Fotolia, P159c: Nidal Yunus, P159d: Matthew Andrews/Dreamstime.com, P159e: Dimitar Bosakov/Shutterstock, P159f: Michael Nivelet/Shutterstock, P159g: Bobby Scrivener/Shutterstock.

Although we have made every effort to trace and contact all copyright holders before publication this has not been possible in all cases. If notified, the publisher will rectify any errors or omissions at the earliest opportunity.

Links to third party websites are provided by Oxford in good faith and for information only. Oxford disclaims any responsibility for the materials contained in any third party website referenced in this work.

The questions, example answers, marks awarded and comments that appear in this book were written by the author(s). In examination, the way marks would be awarded to answers like these may be different.

Contents

Unit 1	**Tens and Ones**	1
	Engage	
	1A Counting in tens and ones	2
	1B Digit values	5
	1C Estimating and counting	7
	Connect	9
	Review	10

Unit 2	**Number Patterns and Properties**	11
	Engage	
	2A Odd and even	12
	2B Doubles	14
	2C Ordering and between	16
	2D Less than, greater than	18
	2E Ordinal numbers	20
	Connect	22
	Review	23

Unit 3	**Number Pairs**	25
	Engage	
	3A Pairs for 10 and rounding	26
	3B Pairs for 20 and 100	28
	Connect and Review	30
	Review	32

Unit 4	**Calculating – Addition and Subtraction**	33
	Engage	
	4A One more, one less; ten more, ten less	34
	4B Adding small numbers	38
	4C Add and subtract a single-digit number	40
	4D Adding two 2-digit numbers	43
	4E Finding the difference	46
	4F Missing numbers	49
	Connect	53
	Review	54

Unit 5	**Number Families**	55
	Engage	
	5A Fact families for number pairs to 20	56
	5B Fact families for multiples of 10 to 100	58
	Connect	60
	Review	61

Unit 6	**Multiplication and Division**	63
	Engage	
	6A Twos	64
	6B Fives and tens	67
	6C Threes and fours	71
	6D Arrays	74
	6E Division as grouping	77
	6F Remainders	81
	Connect	85
	Review	86

Unit 7	Parts of a Whole	89
	Engage	
	7A Half of a shape	90
	7B Half of an amount	92
	7C Quarter and three-quarters of a shape	95
	7D Quarter of an amount	98
	Connect	102
	Review	103

Unit 8	Shapes Everywhere	105
	Engage	
	8A 2D shapes	106
	8B 3D shapes	109
	8C Sorting shapes using a Venn diagram	112
	8D Sorting shapes using a Carroll diagram	114
	8E Reflective symmetry 1	116
	8F Reflective symmetry 2	118
	Connect	120
	Review	121

Unit 9	Measurements	123
	Engage	
	9A Measuring in centimetres	124
	9B String measures	128
	9C Metre measures	131
	9D Liquid measures	134
	9E Sweet cake weights	138
	9F Money	142
	Connect	146
	Review	148

Unit 10	Geometry	149
	Engage	
	10A Turns and right angles	150
	10B Travelling	153
	Connect	156
	Review	157

Unit 11	Time	159
	Engage	
	11A Seconds and minutes	160
	11B Half past	163
	11C Days of the week	166
	11D Months of the year	169
	Connect	173
	Review	174

Unit 12	Handling Data	175
	Engage	
	12A Block graphs and pictograms	176
	12B Sorting using Carroll diagrams	180
	12C Sorting using Venn diagrams	182
	Connect	184
	Review	185

Glossary 187

1 Tens and Ones

Engage

How many newspapers in this pile?

I think there are at least 100.

I guess there are 50 newspapers.

There are seven days in a week. How many newspapers do we buy in a week?

1A Counting in tens and ones

Discover

Counting in ones →

1	2	3	4	5	6	7	8	9	10
11	12	13	14	15	16	17	18	19	20
21	22	23	24	25	26	27	28	29	30
31	32	33	34	35	36	37	38	39	40
41	42	43	44	45	46	47	48	49	50
51	52	53	54	55	56	57	58	59	60
61	62	63	64	65	66	67	68	69	70
71	72	73	74	75	76	77	78	79	80
81	82	83	84	85	86	87	88	89	90
91	92	93	94	95	96	97	98	99	100

Counting in tens ↓

Fill in the missing numbers. Complete each sentence using the words *on* → or *back* ← in the first blank and *ones* or *tens* in the second blank. Always start counting at the first number. The first one has been done for you.

3   4   5   [6]   7   8

This is counting __on__ in __ones__.

15   16   [ ]   18   19   20

This is counting _____ in _____.

[ ]   25   24   [ ]   [ ]   21

This is counting _____ in _____.

17   27   ☐   47   ☐   67

This is counting _____ in _____.

☐   50   40   30   ☐   ☐

This is counting _____ in _____.

☐   79   69   ☐   49   39

This is counting _____ in _____.

☐   ☐   46   45   44   43

This is counting _____ in _____.

☐   12   22   ☐   ☐   ☐

This is counting _____ in _____.

☐   73   63   53   ☐   ☐

This is counting _____ in _____.

32   ☐   34   ☐   36   37

This is counting _____ in _____.

Tens and Ones

3

1A Counting in tens and ones

Explore

1	2	3	4	5	6	7	8	9	10
11	12	13	14	15	16	17	18	19	20
21	22	23	24	25	26	27	28	29	30
31	32	33	34	35	36	37	38	39	40
41	42	43	44	45	46	47	48	49	50
51	52	53	54	55	56	57	58	59	60
61	62	63	64	65	66	67	68	69	70
71	72	73	74	75	76	77	78	79	80
81	82	83	84	85	86	87	88	89	90
91	92	93	94	95	96	97	98	99	100

Colour in 54, 62 and 12 in the same colour on the 100 square. Here is how two students counted from 54 to 62 in ones then back from 62 to 12 in tens.

Start at ☐54☐ , count on in ones to ☐62☐ then count back in tens to ☐12☐ .

Colour in these numbers on the 100 square. ☐67☐ ☐76☐ ☐16☐

Complete each sentence using the three numbers and the words **on** or **back** and **ones** or **tens** in each space.

Start at ☐ , count _____ in _____ to ☐ then count _____ in _____ to ☐ .

Using the same set of numbers, find a different way to complete the sentence.

1B Digit values

Discover

1	2	3	4	5	6	7	8	9	10
11	12	13	14	15	16	17	18	19	20
21	22	23	24	25	26	27	28	29	30
31	32	33	34	35	36	37	38	39	40
41	42	43	44	45	46	47	48	49	50
51	52	53	54	55	56	57	58	59	60
61	62	63	64	65	66	67	68	69	70
71	72	73	74	75	76	77	78	79	80
81	82	83	84	85	86	87	88	89	90
91	92	93	94	95	96	97	98	99	100

Which place value cards do you need to make these numbers? The first one is done for you.

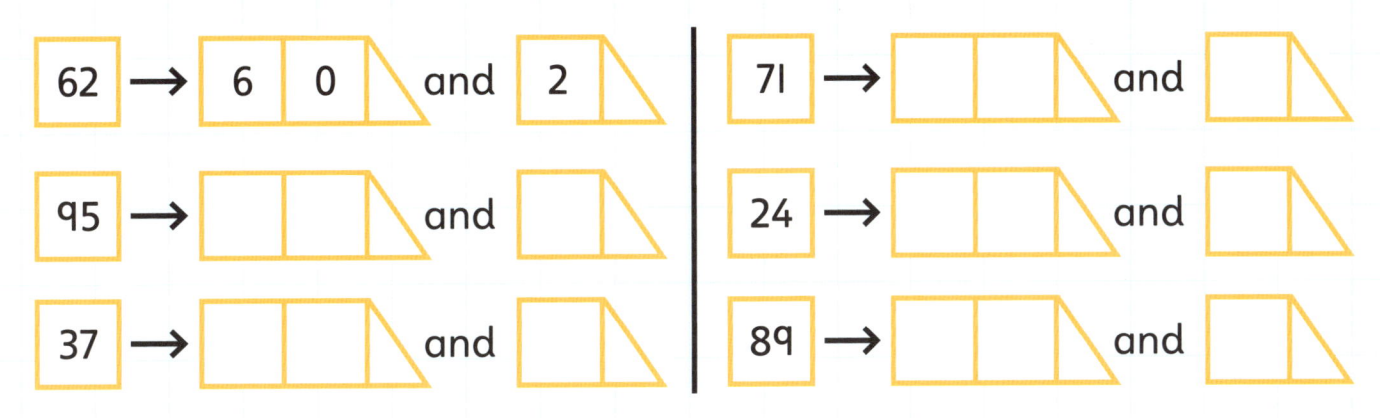

62 → 6 0 and 2

95 → ☐ and ☐

37 → ☐ and ☐

71 → ☐ and ☐

24 → ☐ and ☐

89 → ☐ and ☐

Colour in these numbers on the 100 square. From a set of place value cards, take out all the cards used above.

Which numbers are left? _____

Make as many different 2-digit numbers as you can using the cards that are left.

Tens and Ones

1B Digit values

Explore

Complete the value of each set of place value counters.
The first one has been done for you.

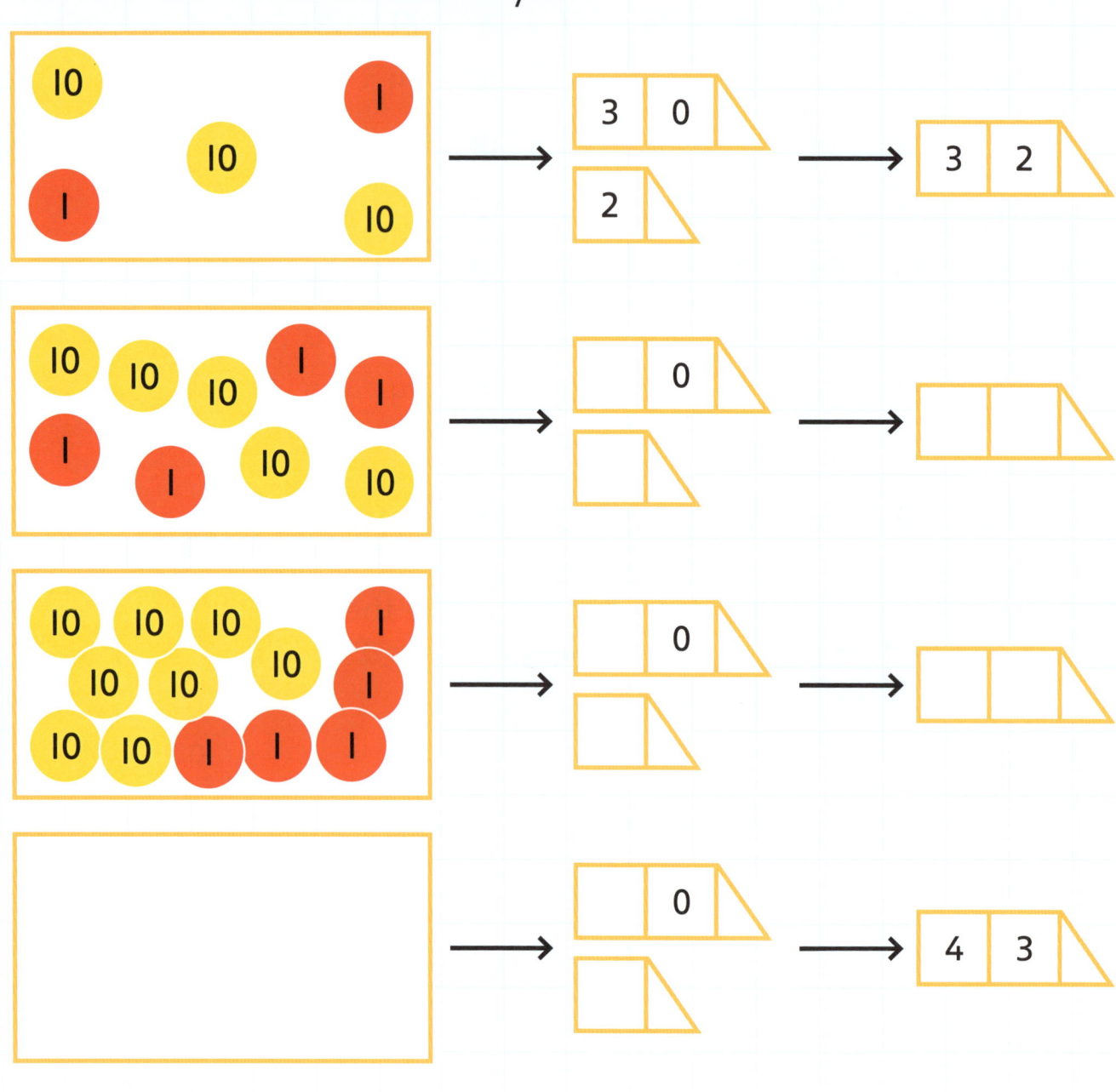

Box 1: three "10" counters and two "1" counters → 30 / 2 → 32

Box 2: five "10" counters and four "1" counters → _0 / _ → __

Box 3: eight "10" counters and five "1" counters → _0 / _ → __

Box 4: (empty) → _0 / _ → 43

Box 5: (empty) → _0 / _ → 71

6

1C Estimating and counting

Discover

Your teacher will give you some bags of objects. Estimate how many objects are in each bag. Tip out the bag and count the objects. Label each bag so you know which objects you counted.

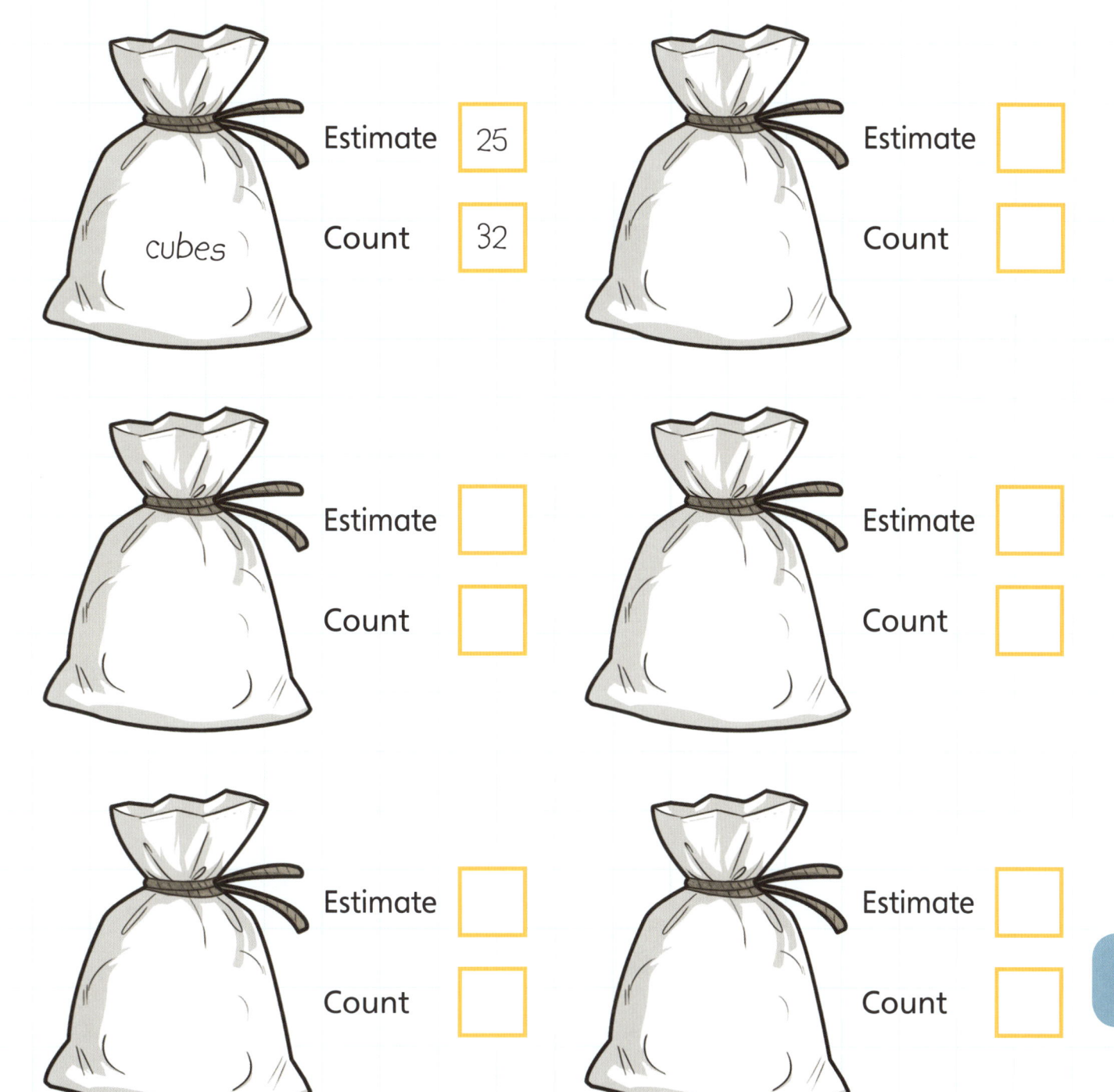

Bag labeled "cubes" — Estimate: 25, Count: 32

Estimate: ☐ Count: ☐

Estimate: ☐ Count: ☐

Estimate: ☐ Count: ☐

Estimate: ☐ Count: ☐

Estimate: ☐ Count: ☐

Tens and Ones

7

1C Estimating and counting

Explore

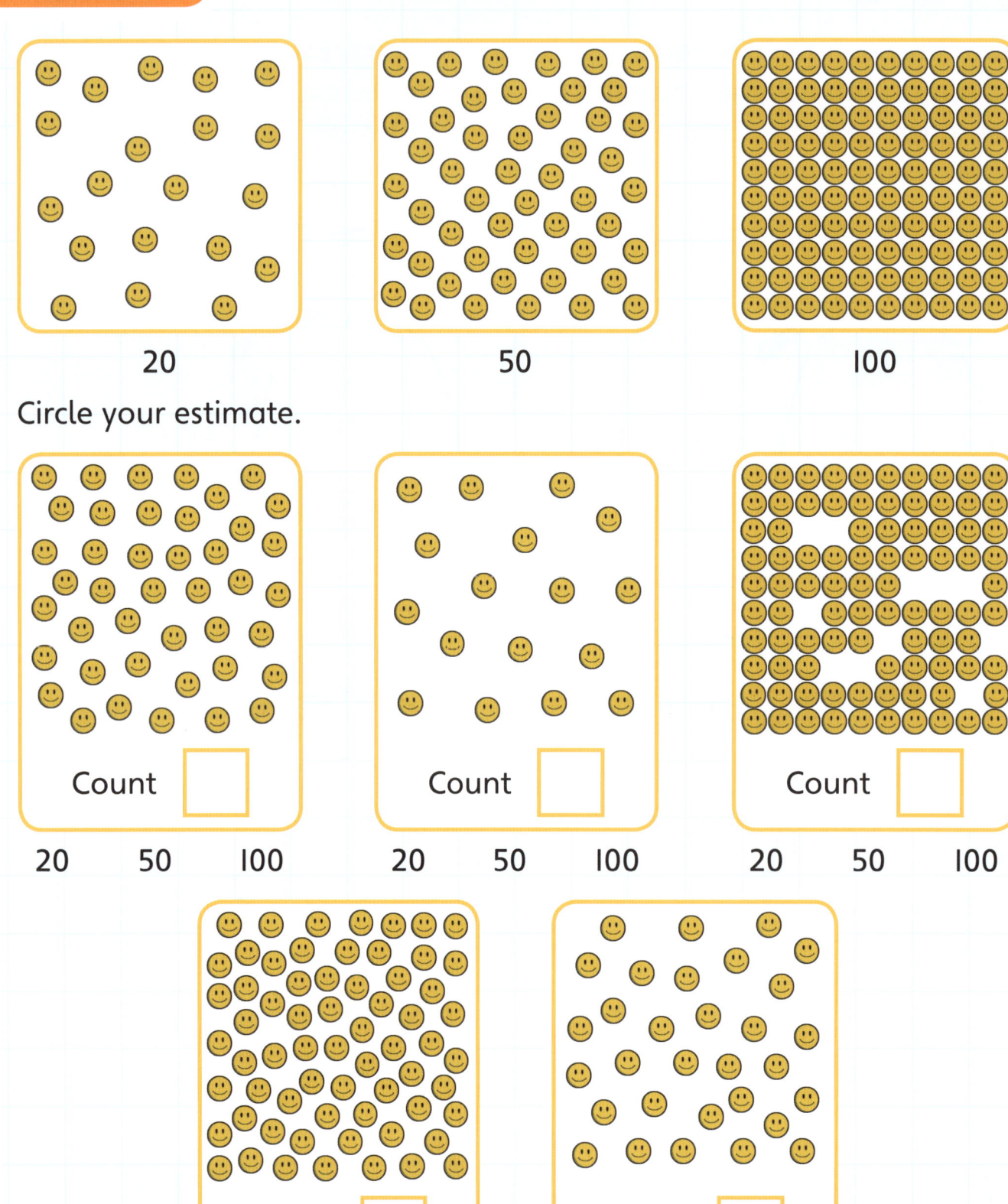

20   50   100

Circle your estimate.

Count ☐   Count ☐   Count ☐
20 50 100   20 50 100   20 50 100

Count ☐   Count ☐
20 50 100   20 50 100

1 Tens and ones

Connect

You have a bag of 9 counters, a mixture of yellow counters with a value of 10 and red counters with a value of 1.

1   10

Which numbers could you make with your place value counters? You do not have to use all the counters when you make a number.

Colour the numbers you can make on the 100 square.

1	2	3	4	5	6	7	8	9	10
11	12	13	14	15	16	17	18	19	20
21	22	23	24	25	26	27	28	29	30
31	32	33	34	35	36	37	38	39	40
41	42	43	44	45	46	47	48	49	50
51	52	53	54	55	56	57	58	59	60
61	62	63	64	65	66	67	68	69	70
71	72	73	74	75	76	77	78	79	80
81	82	83	84	85	86	87	88	89	90
91	92	93	94	95	96	97	98	99	100

Which numbers was it **not** possible to make?

Tell a friend why you cannot make these numbers.

I can't make _____ because _____.

1 Tens and ones

Review

Use the clues to help you find the numbers.

I have 2 tens and 4 ones. I am ☐

I have 6 tens and 8 ones. I am ☐

I have 3 ones and 7 tens. I am ☐

I have 9 ones and 5 tens. I am ☐

Draw the matching counters for each number. ● 1   ● 10

26

43

2 Number Patterns and Properties

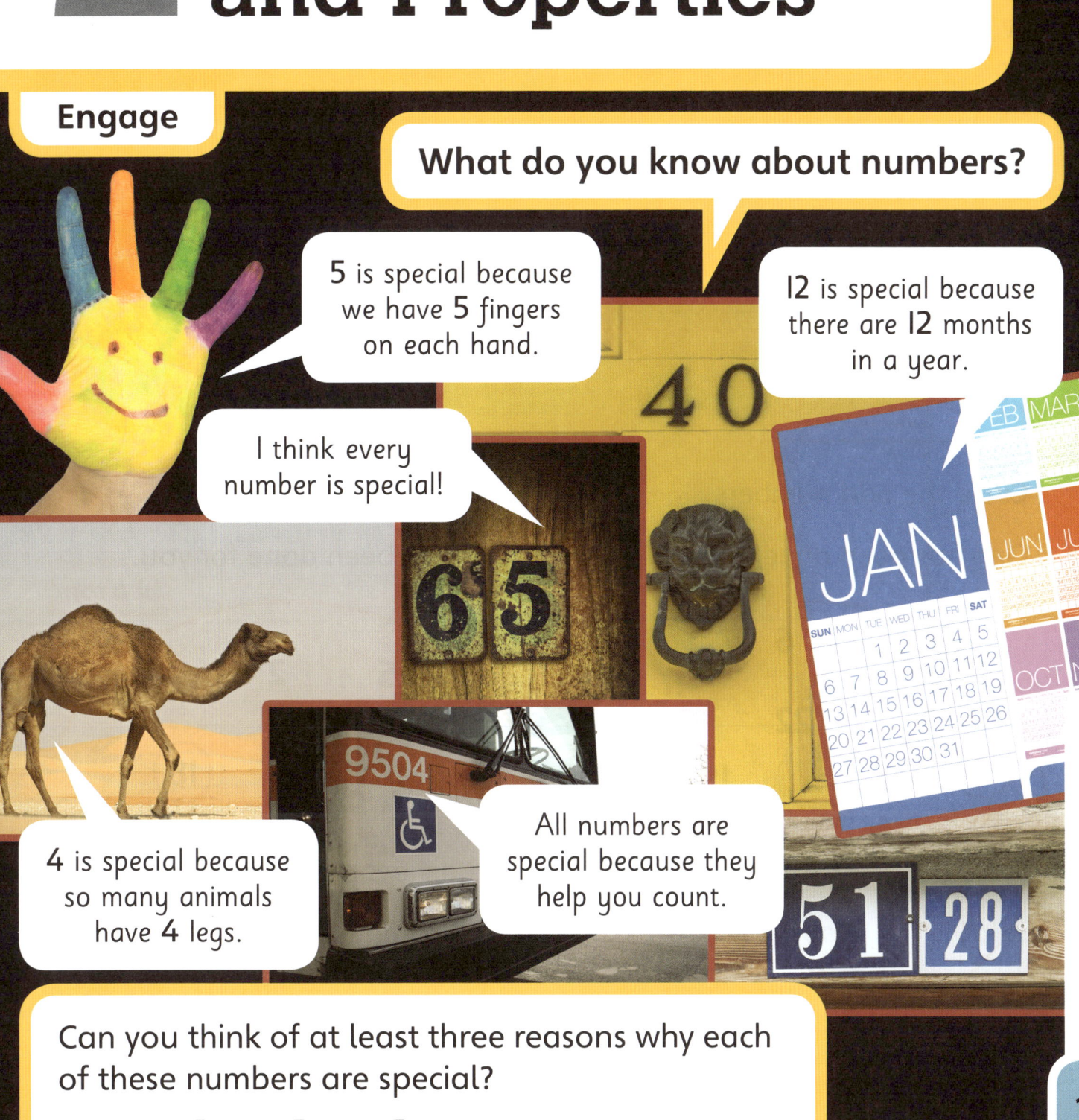

Engage

What do you know about numbers?

5 is special because we have 5 fingers on each hand.

12 is special because there are 12 months in a year.

I think every number is special!

4 is special because so many animals have 4 legs.

All numbers are special because they help you count.

Can you think of at least three reasons why each of these numbers are special?

7   2   5   1

Choose a number. Explain why it is special to you.

2B Doubles

Discover

Complete the dominoes and work out the doubles.

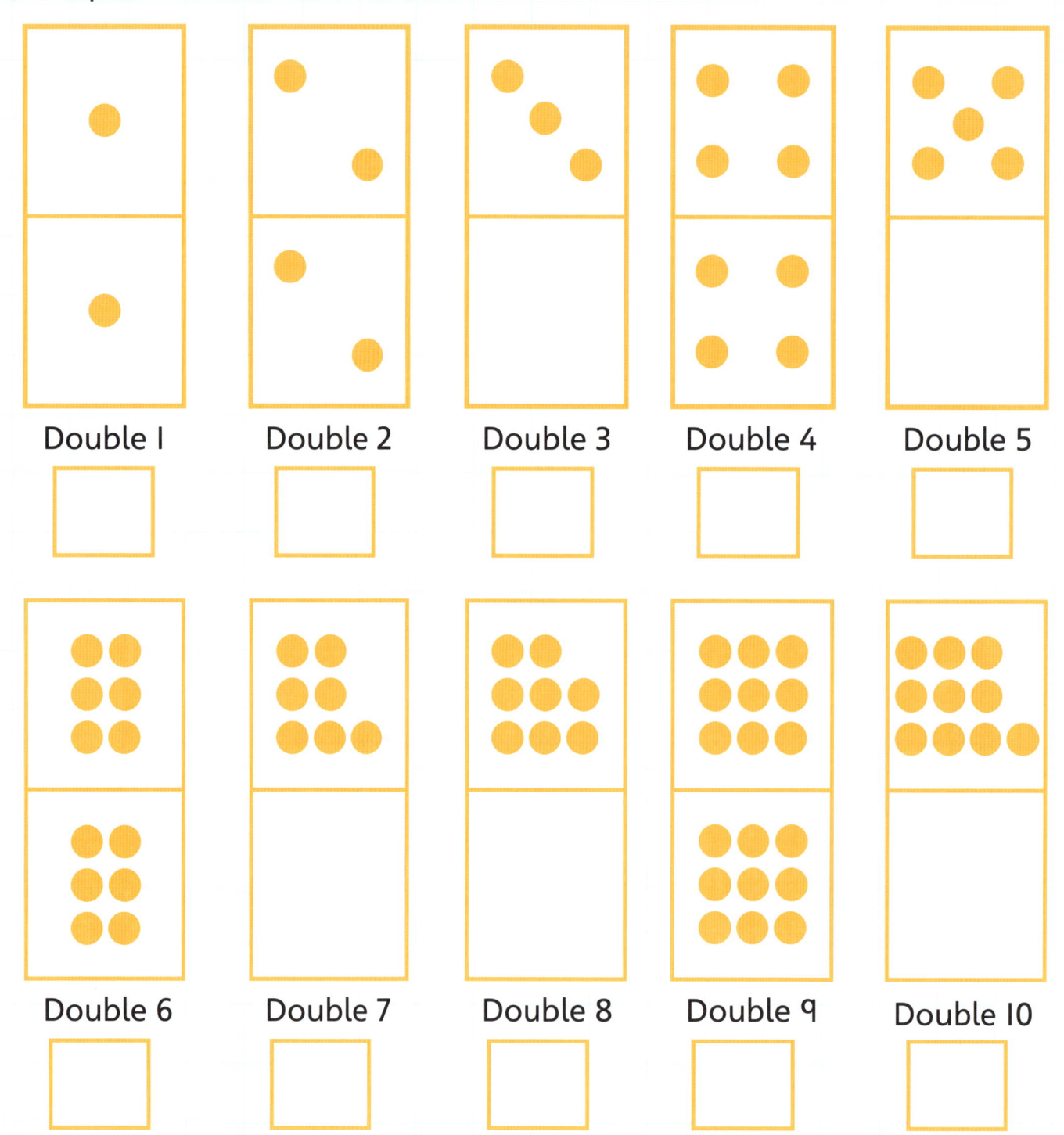

Double 1   Double 2   Double 3   Double 4   Double 5

Double 6   Double 7   Double 8   Double 9   Double 10

Complete the sentence with the word **odd** or **even**.

Doubles of the numbers 1 to 10 are all _____ numbers.

2B Doubles

Explore

Use the doubles you already know to work out these doubles.

12   14   17

Split into tens and ones.

10   2

Double each part.

20   4

Add the doubled parts together.

24

Now work out the doubles for each of these numbers.

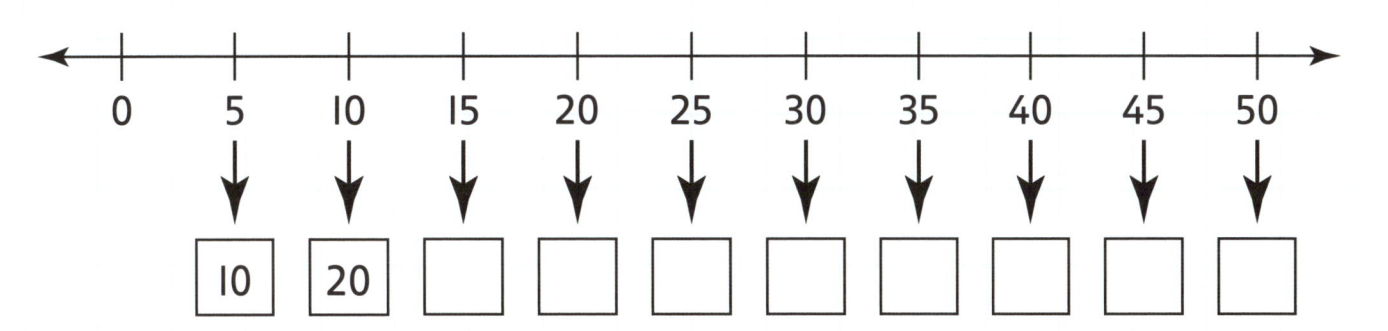

0   5   10   15   20   25   30   35   40   45   50

10 20

Complete this sentence using 0 or 5.

When you double a number with 5 or 0 ones,

its double always has _____ ones.

Number Patterns and Properties

2C Ordering and between

Discover

Write a number that is between each pair of tens numbers.

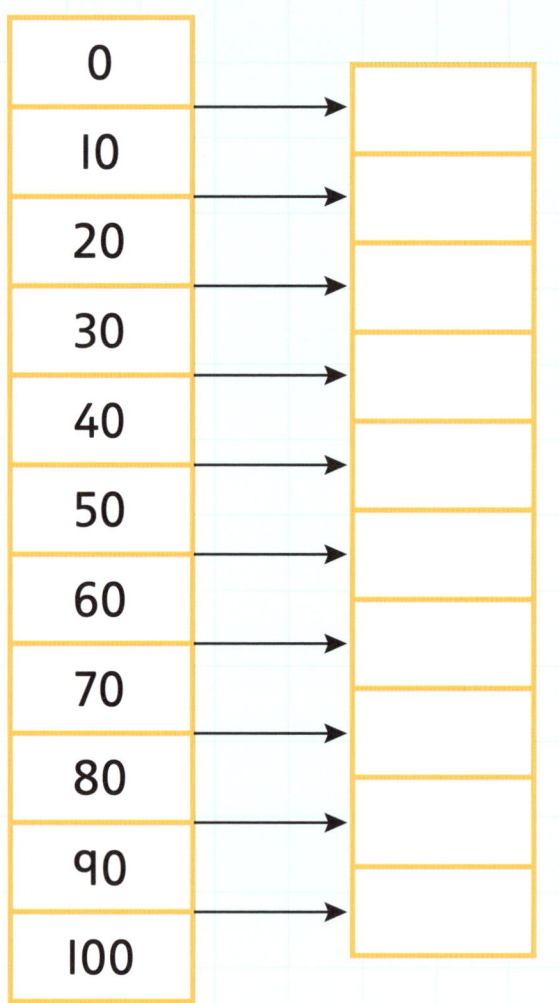

| 0 |
| 10 |
| 20 |
| 30 |
| 40 |
| 50 |
| 60 |
| 70 |
| 80 |
| 90 |
| 100 |

between

Between means more than one number but less than another.

Write one or two numbers between.
The first one has been done for you.

| 15 | 16 | 17 |

| 34 | | 36 |

| 9 | | 11 |

| 12 | | | 18 |

| 39 | | | 43 |

| 28 | | 30 |

| 60 | | | 70 |

| 40 | | | 50 |

| 10 | | | 20 |

16

2C Ordering and between

Explore

Cross out the numbers you have made on this square.

Colour in the numbers between the numbers that you crossed out. Use a different colour for each set of numbers.

Write the colour you used for each set inside the number square.

1	2	3	4	5	6	7	8	9	10
11	12	13	14	15	16	17	18	19	20
21	22	23	24	25	26	27	28	29	30
31	32	33	34	35	36	37	38	39	40
41	42	43	44	45	46	47	48	49	50
51	52	53	54	55	56	57	58	59	60
61	62	63	64	65	66	67	68	69	70
71	72	73	74	75	76	77	78	79	80
81	82	83	84	85	86	87	88	89	90
91	92	93	94	95	96	97	98	99	100

I coloured the numbers between _____ and _____

I coloured the numbers between _____ and _____

I coloured the numbers between _____ and _____

I coloured the numbers between _____ and _____

I coloured the numbers between _____ and _____

I coloured the numbers between _____ and _____

I coloured the numbers between _____ and _____

2D Less than, greater than

Discover

Look at number 8.

$$4 < 8$$

This means that 4 **is less than** 8

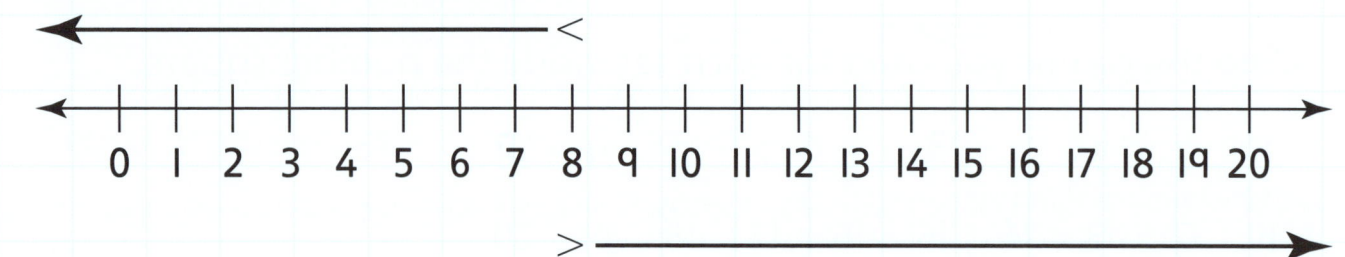

←——————————<

+—+→
0 1 2 3 4 5 6 7 8 9 10 11 12 13 14 15 16 17 18 19 20

>——————————————————→

$$16 > 8$$

This means that 16 **is greater than** 8

<	>
is less than	is greater than

+—+→
0 1 2 3 4 5 6 7 8 9 10 11 12 13 14 15 16 17 18 19 20

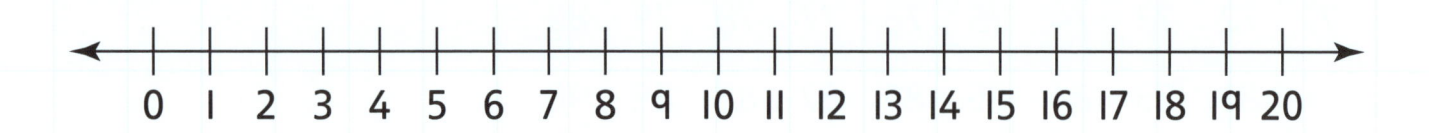

We use the signs < and > to compare numbers.

Complete these sentences using a number or < or >.
Use the number line to help you.

9 < ◯   15 > ◯   10 ◯ 14

18 ◯ 12   6 ◯ 2   11 ◯ 19

17 ◯ 13   16 < ◯   ◯ > 10

18

2D Less than, greater than

Explore

Use place value cards to make nine 2-digit numbers.
Put the numbers in order, from smallest to largest.
Write your numbers in the grid.

Here is an example of a completed grid.

| 14 | 23 | 38 | 41 | 56 | 62 | 75 | 89 | 97 |

| | | | | | | | | |

<
is less than

\>
is greater than

Write some number sentences to compare your numbers.

☐ < ☐   ☐ > ☐

☐ < ☐   ☐ > ☐

☐ < ☐   ☐ > ☐

☐ < ☐   ☐ > ☐

☐ < ☐   ☐ > ☐

☐ < ☐   ☐ > ☐

2E Ordinal numbers

Discover

1	first	1st
2	second	2nd
3	third	3rd
4	fourth	4th
5	fifth	5th

6	sixth	6th
7	seventh	7th
8	eighth	8th
9	ninth	9th
10	tenth	10th

Colour the 1st bead red.

Colour the 2nd bead blue.

Colour the 5th bead red.

Colour the 3rd bead red.

Colour the 6th bead blue.

Colour the 8th bead yellow.

Colour the 4th bead yellow.

Colour the 7th bead red.

Continue the pattern.

What colour is the 12th bead? ◯

What colour is the 15th bead? ◯

What colour is the 16th bead? ◯

What colour is the 19th bead? ◯

Answer the following questions.

What is the 4th letter of your name? _____

What is the 3rd odd number? _____

What is the next number after 20? _____

20

2E Ordinal numbers

Explore

Monday	Tuesday	Wednesday	Thursday	Friday	Saturday	Sunday
1	2	3	4	5	6	7
8	9	10	11	12	13	14
15	16	17	18	19	20	21
22	23	24	25	26	27	28
29	30	31				

Answer these questions and complete these sentences about the calendar.

The first day of the month is a _____.

The first Friday of the month is the _____.

Which day of the week is the 11th? _____

Which day of the week is the 6th? _____

Which day of the week is the 16th? _____

Which day of the week is the 2nd? _____

h	f	e	n
t	a	l	p
u	s	w	i
y	d	o	r

Use these clues to find the words

1st letter on the 4th line.
3rd letter on the 1st line.
2nd letter on the 3rd line.
_____ _____ _____

1st letter on the 2nd line.
3rd letter on the 3rd line.
3rd letter on the 4th line.
_____ _____ _____

1st letter on the 2nd line.
4th letter on the 4th line.
1st letter on the 3rd line.
3rd letter on the 1st line.
_____ _____ _____ _____

Number Patterns and Properties

21

2 Number patterns and properties

Connect

Complete each spider diagram with as much information as you can about the number. Then choose a number to explore.

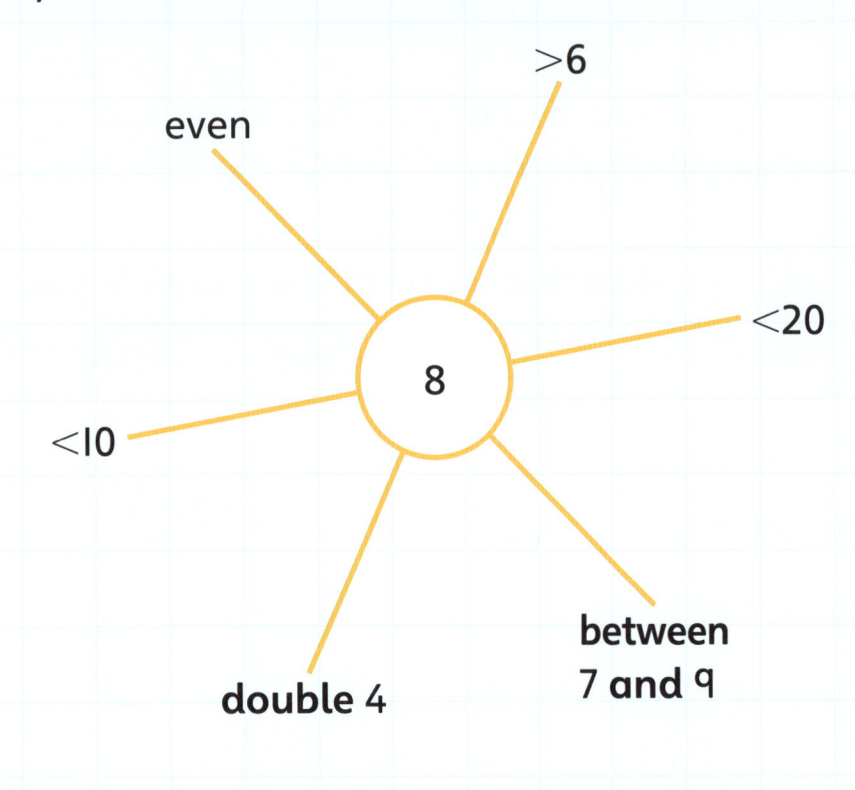

- even
- >6
- <20
- between 7 and 9
- double 4
- <10

8

17

24

2 Number patterns and properties

Review

Complete these sentences using odd or even.

Numbers with 0, 2, 4, 6 or 8 ones are _____.

Numbers with 1, 3, 5, 7 or 9 ones are _____.

These numbers have been ordered from largest to smallest. Write a number between each pair of numbers. Make sure the numbers are still in order from largest to smallest.

Double these numbers.

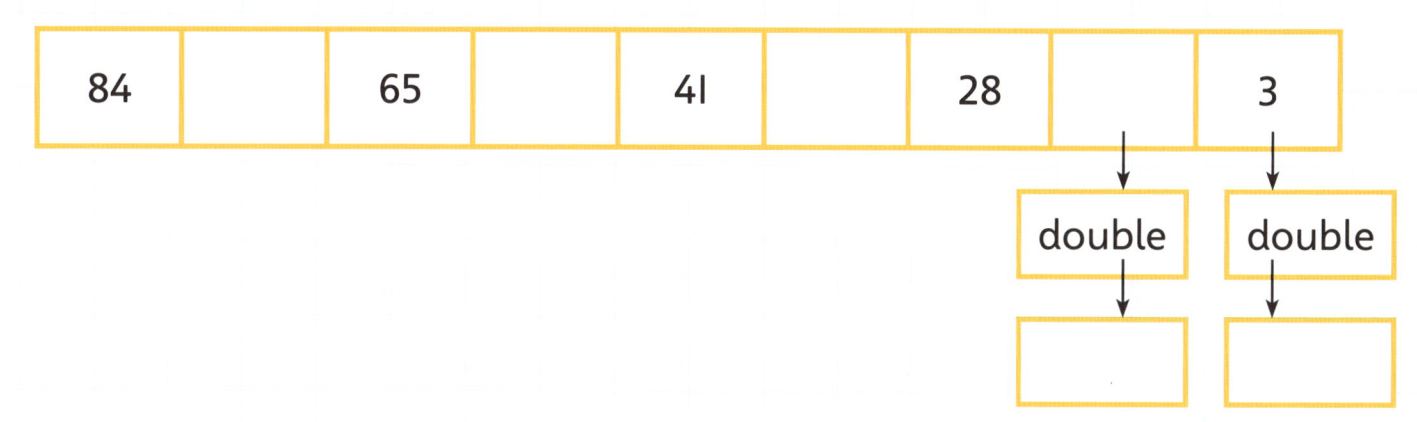

| 84 | | 65 | | 41 | | 28 | | 3 |

double   double

Write the fourth number here. ☐

Use any of the numbers above to complete these number sentences.

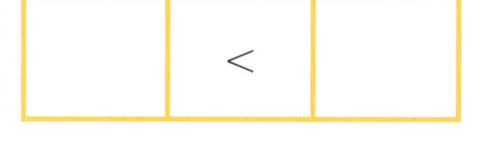

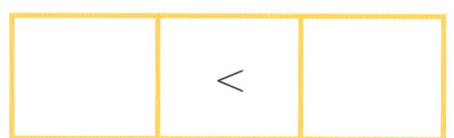

☐ < ☐   ☐ < ☐

☐ < ☐   ☐ < ☐

☐ < ☐   ☐ < ☐

23

3 Number Pairs

Engage

What are number pairs?

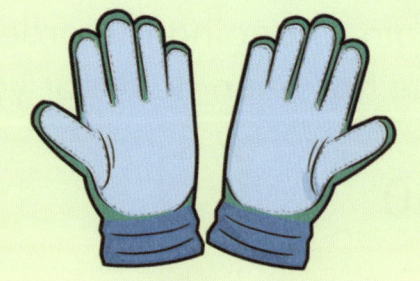

Pair means two things that are the same.

A pair of shoes are not exactly the same. There's a left foot and a right foot.

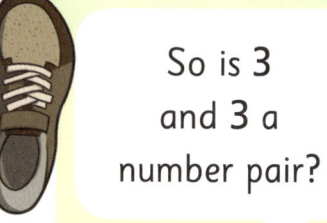

So is 3 and 3 a number pair?

3 and 3 makes 6 but so does 4 and 2.

Number Pairs

25

3A Pairs for 10 and rounding

Discover

| ~~0~~ | 1 | 2 | 3 | 4 | 5 | 6 | 7 | 8 | 9 | ~~10~~ |

Use the numbers from 0 to 10 to find all the number pairs for 10.

You can only use each number once. Cross out each number as you use it. The first one has been done for you.

Then write each pair again, changing the order of the numbers in each row. The first one has been done for you.

| 0 + 10 = 10 |
| |
| |
| |
| |

| 10 + 0 = 10 |
| |
| |
| |
| |

> It does not matter which order you write the numbers in, they are the same pair.

Did you use all the numbers? Which number is left? ☐

Write the number pair for 10 using this number.   ☐ + ☐ = ☐

How many **different** number pairs for 10 are there? ☐

26

3A Pairs for 10 and rounding

Explore

round down | round up

0	1	2	3	4	5	6	7	8	9	10
	11	12	13	14	15	16	17	18	19	20
	21	22	23	24	25	26	27	28	29	30
	31	32	33	34	35	36	37	38	39	40
	41	42	43	44	45	46	47	48	49	50
	51	52	53	54	55	56	57	58	59	60
	61	62	63	64	65	66	67	68	69	70
	71	72	73	74	75	76	77	78	79	80
	81	82	83	84	85	86	87	88	89	90
	91	92	93	94	95	96	97	98	99	100

Fill in the tens numbers in the first column of the 100 square.

Use the completed 100 square to help you round these numbers to the nearest 10.

62 ☐   89 ☐   25 ☐

55 ☐   37 ☐   9 ☐

3 ☐   46 ☐   95 ☐

31 ☐   20 ☐   5 ☐

Complete the sentences using the word **up** or **down**.

If the ones digit is 4 or less, you round _____.

If the ones digit is 5 or more, you round _____.

Number Pairs

27

3B Pairs for 20 and 100

Discover

Here are all the multiples of 10 to 100. Use these numbers to find all the number pairs for 100 using multiples of 10.

| ~~0~~ | 10 | 20 | 30 | 40 | 50 | 60 | 70 | 80 | 90 | ~~100~~ |

You can only use each number once. Cross out each number as you use it. The first one has been done for you.

Then write each pair again, changing the order of the numbers. The first one has been done for you.

0 + 100 = 100	100 + 0 = 100

It does not matter which order you write the numbers in, they are the same pair.

Did you use all the numbers? Which number is left? ☐

Write the number pair for 100 using this number.   ☐ + ☐ = ☐

Complete the sentence:

Finding number pairs for 100 using multiples of 10 is the same as finding number pairs for 10 except the numbers are _____ times bigger.

3B Pairs for 20 and 100

Explore

You have found number pairs for 10 and 100, now find the pairs for 20.

0	1	2	3	4	5	6	7	8	9	10
11	12	13	14	15	16	17	18	19	20

You can only use each number once. Cross out each number as you use it. The first one has been done for you.

0 + 20 = 20

Make sure you have not written the same pair twice but in a different order.

Did you use all the numbers? Which number is left?

Write the number pair for 20 using this number.

☐ + ☐ = ☐

Imagine you are finding number pairs for 30.
Would you use all the numbers?

Which number would not be used?

How many number pairs do you think there are for 30?

Number Pairs

29

3 Number pairs

Connect

You will need:

- squared paper
- counters.

I like to make cakes for my friends. Some people like chocolate cakes, some people like lemon cakes and some people like both chocolate cakes and lemon cakes. I have boxes which hold 12, 14, 16, 18 or 20 cakes.

Here is a box of 12 chocolate cakes.

Some of my boxes were damaged. I cannot put a cake in one corner.

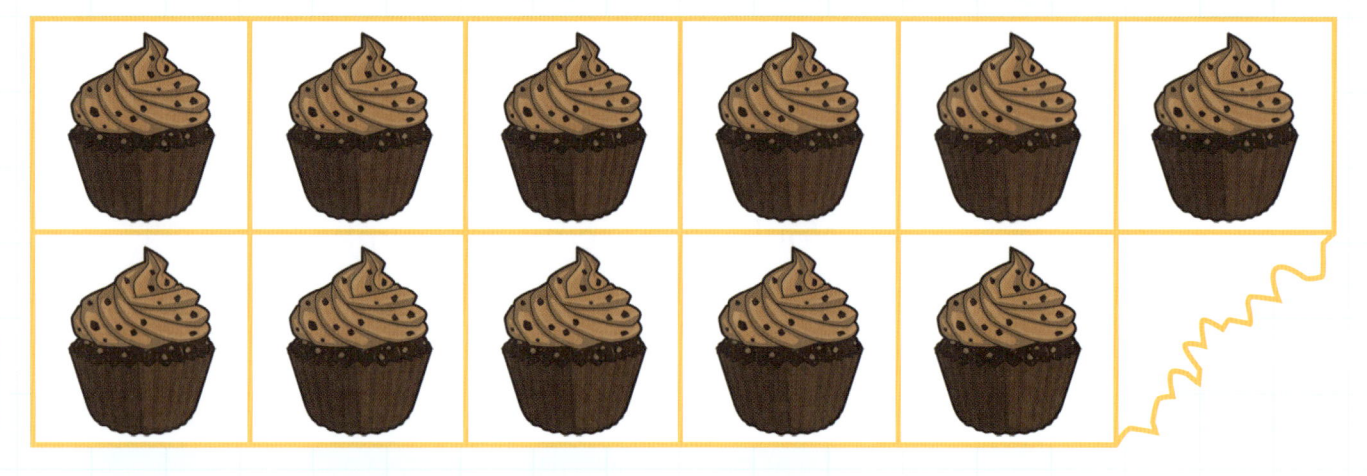

This box for 12 cakes will only hold 11 cakes.
Each of the damaged boxes hold one cake less.

In your group, choose one size of box and write all the different ways of filling the box with chocolate and lemon cakes.

How will you record what you find out?
You might find squared paper and counters useful.

Example:

14 + 6 = 20

14 + 6 = 20

> The boxes are different! Perhaps we should always write the number of chocolate cakes first?

Find all the number pairs for 13.

3 Number Pairs

Review

Write a number to make the sentence or number sentence correct.

_____ + 7 = 10

40 + _____ = 100

2 + _____ + 10

_____ is 40 rounded to the nearest 10

_____ is 60 rounded to the nearest 10

14 + _____ = 20

_____ + 8 = 20

_____ is 90 rounded to the nearest 10

_____ + 10 = 100

11 + _____ = 20

_____ is 50 rounded to the nearest 10

Number Facts

Write 5 different facts about the number 13

Pick any two-digit number: Write 5 facts that you know about your number.

4 Calculating – Addition and Subtraction

Engage

How do you add two numbers together?

I just count on in my head.

I use a number line.

I split the tens and ones.

It depends on what the numbers are.

4A One more, one less; ten more, ten less

Discover

I more →
I less ←

1	2	3	4	5	6	7	8	9	10
11	12	13	14	15	16	17	18	19	20
21	22	23	24	25	26	27	28	29	30
31	32	33	34	35	36	37	38	39	40
41	42	43	44	45	46	47	48	49	50
51	52	53	54	55	56	57	58	59	60
61	62	63	64	65	66	67	68	69	70
71	72	73	74	75	76	77	78	79	80
81	82	83	84	85	86	87	88	89	90
91	92	93	94	95	96	97	98	99	100

10 more/less

Choose a number on the 100-square, then write the number that is **1 more** and **1 less**. Colour each set of numbers a different colour. See the example in the 100-square.

☐ I less 47 I more ☐

☐ I less ☐ I more ☐

☐ I less ☐ I more ☐

☐ I less ☐ I more ☐

34

Choose a number on the 100-square, then write the number that is 10 more and 10 less. Colour each set of numbers a different colour. See the example in the 100-square.

	10 less	73	10 more	
	10 less		10 more	
	10 less		10 more	
	10 less		10 more	
	10 less		10 more	
	10 less		10 more	

Tick ✓ the box if the sentence is correct.

You go right on the 100-square to find one more. ☐

You move down on the 100-square to find 10 more. ☐

You move left on the 100-square to find one less. ☐

You move up on the 100-square to find 10 less. ☐

4A One more, one less; ten more, ten less

Explore

Fill in the missing numbers on these caterpillars. Use a 100-square to help you.

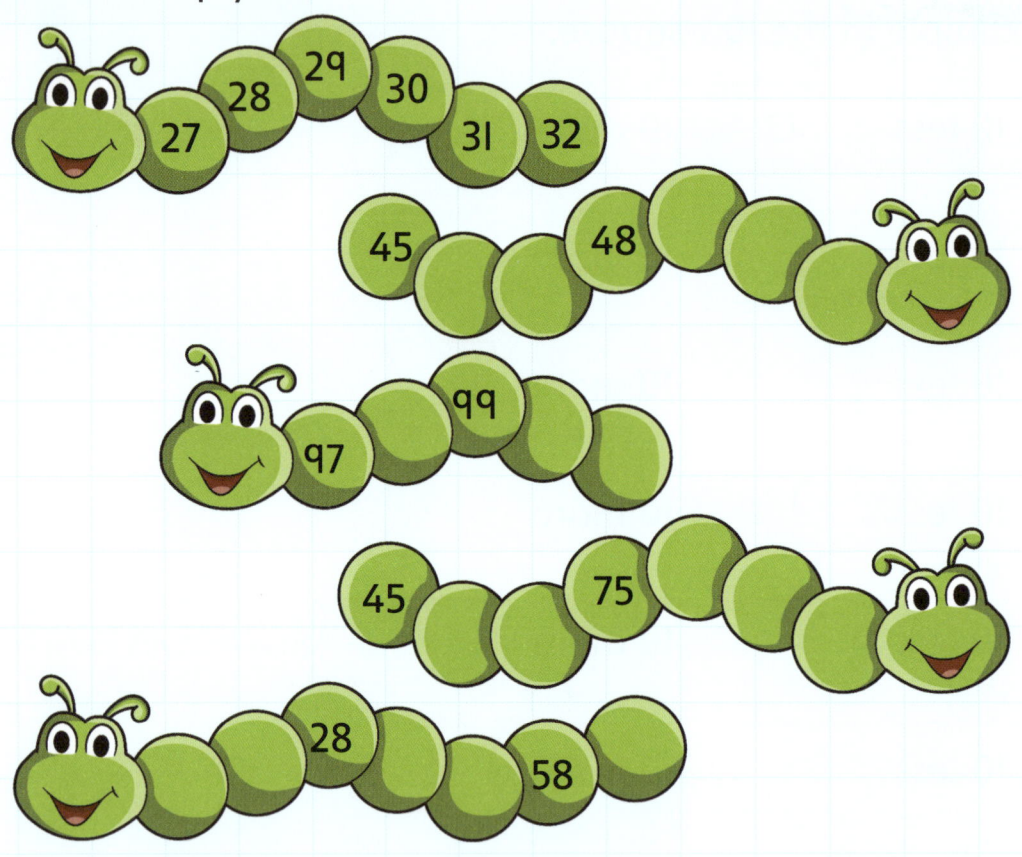

27 28 29 30 31 32

45 48

97 99

45 75

28 58

Write 1 more and 1 less than each number.

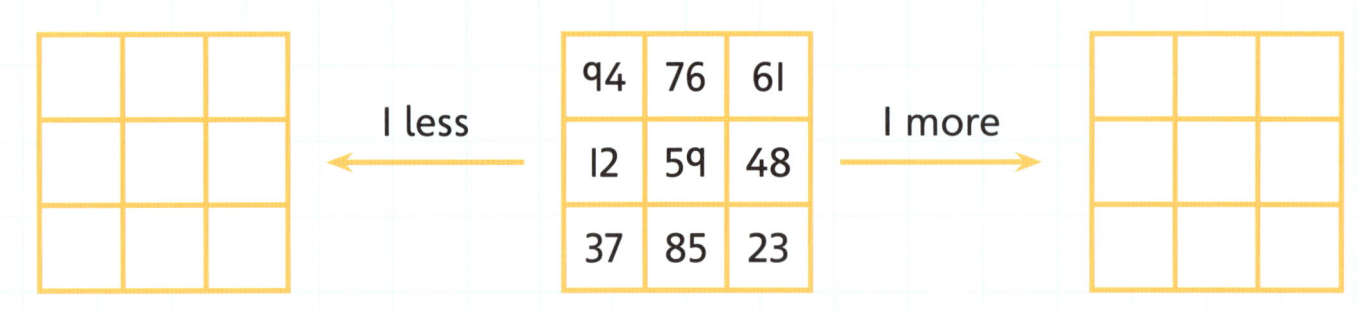

94	76	61
12	59	48
37	85	23

← 1 less   1 more →

Write 10 more and 10 less than each number.

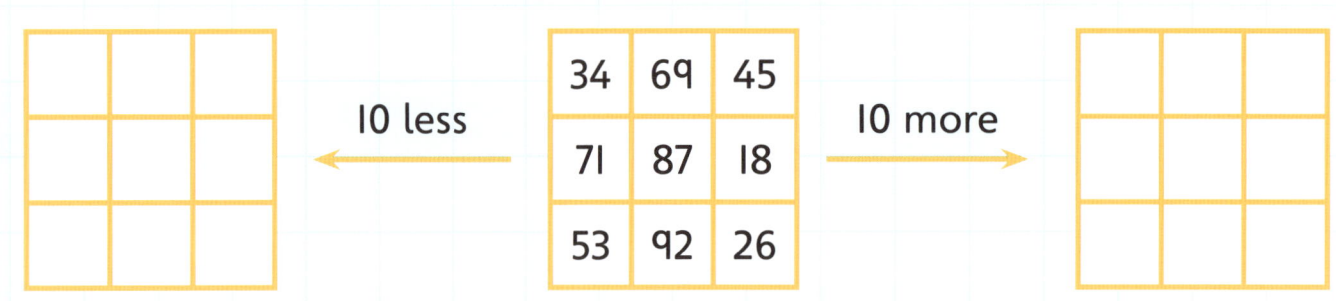

34	69	45
71	87	18
53	92	26

← 10 less   10 more →

Which number is left?

Numbers on caterpillar: 29, 36, 25, 32, 24, 51

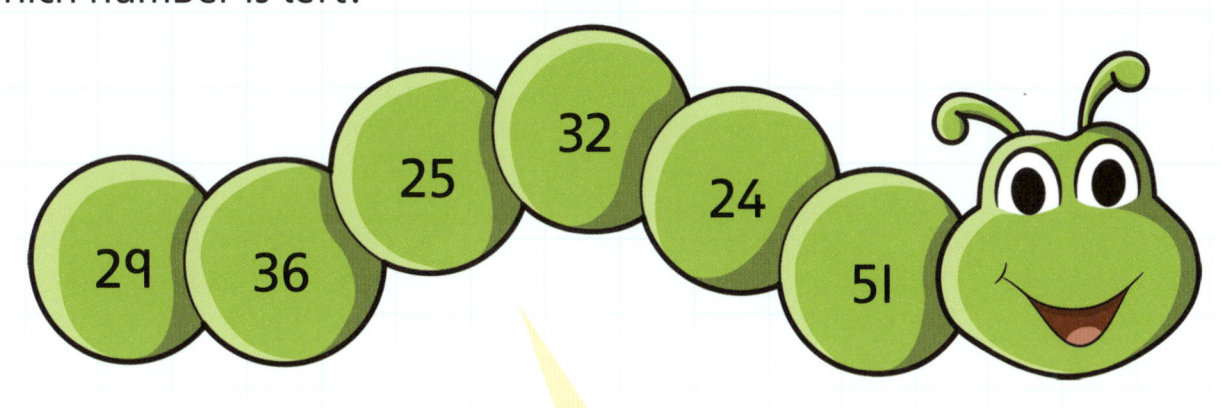

I am 1 more than 23. ☐   I am 1 more than 28. ☐

I am 1 less than 37. ☐   I am 1 more than 31. ☐

I am 1 less than 52. ☐   The number left is _____.

Complete these sentences with the word **ones** or **tens**.
The first one has been done for you.

When you write the number that is 10 or 20 more, the __ones__ digit stays the same.

When you write the number that is 10 or 20 less, the _____ digit stays the same.

When you write the number that is 1 less, the _____ digit usually stays the same.

When you write the number that is 1 more, the _____ digit usually stays the same.

4B Adding small numbers

Discover

Write the six number pairs for 10.

Use the number pairs to help you add these numbers.
The first one has been done for you.

     10
(4) + 3 + (6) = 13

1 + 6 + 9 = ☐

8 + 4 + 6 = ☐

5 + 7 + 5 = ☐

8 + 5 + 2 = ☐

6 + 3 + 7 = ☐

5 + 5 + 4 = ☐

4 + 5 + 6 = ☐

Now add these numbers together.
Look for number pairs for 10 or three numbers to make 10.

1 + 2 + 3 + 4 + 5 = ☐

2 + 3 + 4 + 5 + 6 = ☐

2 + 2 + 6 + 6 = ☐

7 + 6 + 5 + 4 + 3 = ☐

4 + 5 + 6 + 7 = ☐

3 + 4 + 3 + 4 = ☐

8 + 9 + 1 + 1 = ☐

2 + 4 + 6 + 8 = ☐

4B Adding small numbers

Explore

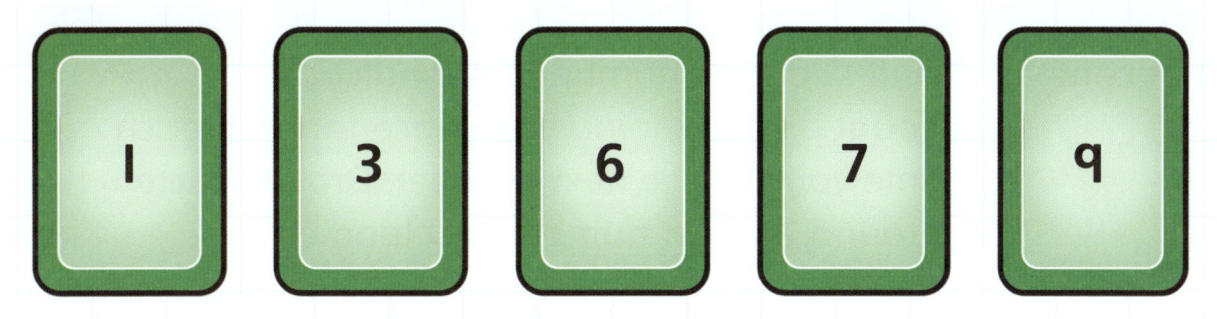

| 1 | 3 | 6 | 7 | 9 |

Choose four of these numbers. Add them together.
How many **different** totals can you make?

Add all five of the numbers together.
Put them in a different order to help you.

_____ + _____ + _____ + _____ + _____ = _____

Remove 0 from a set of 0–9 digit cards. Shuffle and turn over the top five cards. Add them together. Put the cards in a different order to help you add them. Shuffle the cards and turn over the top five again.

_____ + _____ + _____ + _____ + _____ = _____

_____ + _____ + _____ + _____ + _____ = _____

_____ + _____ + _____ + _____ + _____ = _____

_____ + _____ + _____ + _____ + _____ = _____

4C Add and subtract a single-digit number

Discover

Worked example

12 + 7 = 19

Find 12 on the number line and circle it. Count on 7, jumping one space forward each time you say the next number. Draw a circle around the last number you land on.

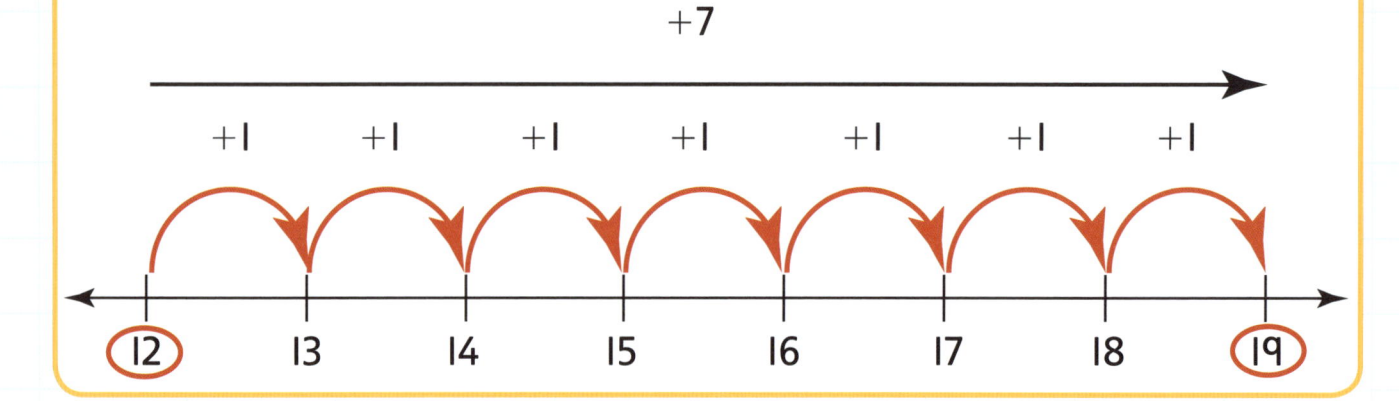

13 + 4 =

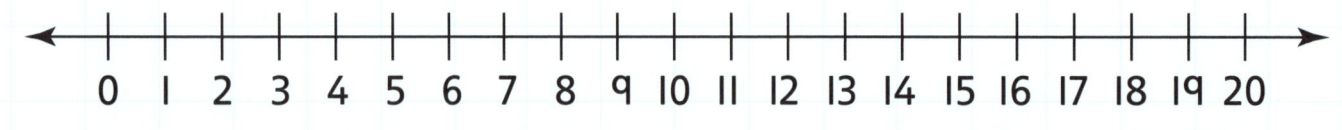

11 + 8 =

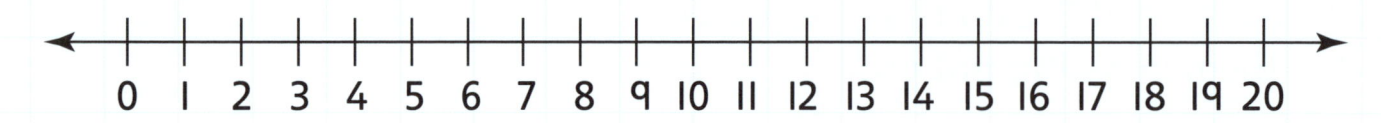

14 − 6 =

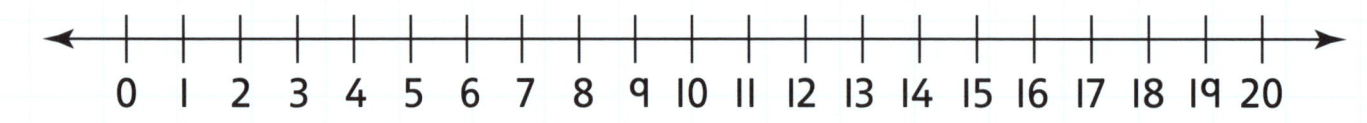

17 − 8 =

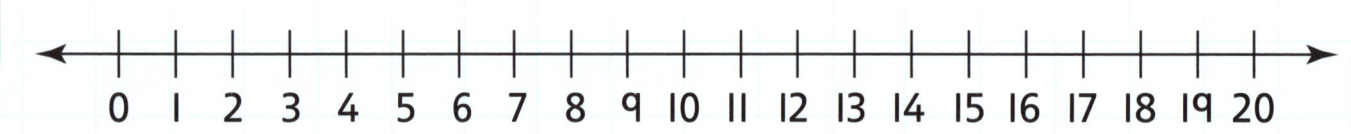

to/from a 2-digit number

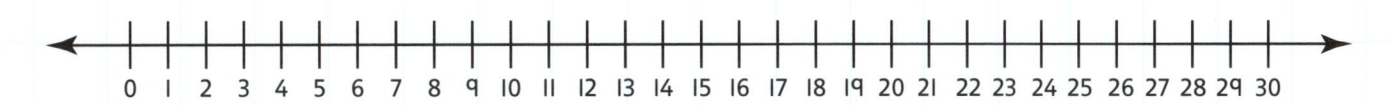

16 + 9 =

22 + 6 =

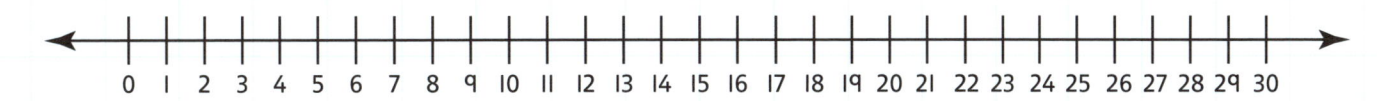

29 − 5 =

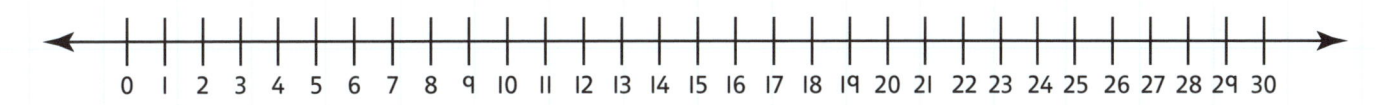

22 − 7 =

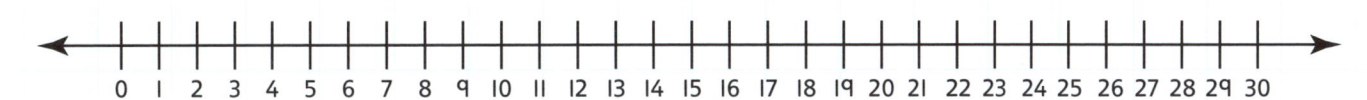

28 − 9 =

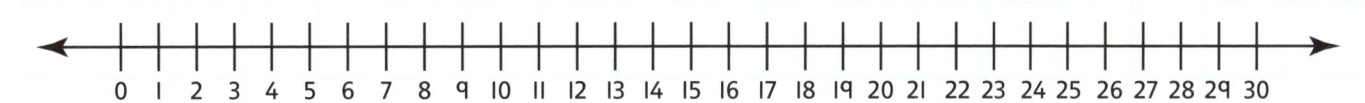

Complete these sentences using the word **right** or **left**.

When you **add** on a number line, you count along the number line to the _____.

When you **take away** on a number line, you count back on the number line to the _____.

4C Add and subtract a single-digit number

Explore

Add or subtract to find the missing numbers.
Use the number line to help you.

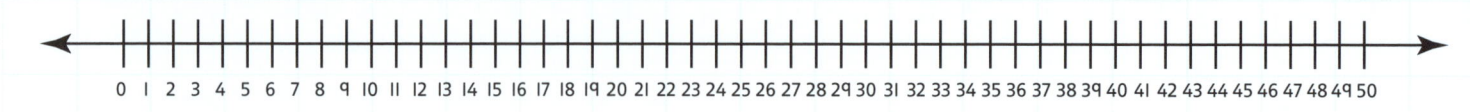

0 1 2 3 4 5 6 7 8 9 10 11 12 13 14 15 16 17 18 19 20 21 22 23 24 25 26 27 28 29 30 31 32 33 34 35 36 37 38 39 40 41 42 43 44 45 46 47 48 49 50

Subtract 8	
19	
21	
47	
26	
32	

+	4	7	9
17			
24			
32			

Add 6	
19	
21	
47	
26	
32	

Add 8	
36	
27	
14	
21	
18	

−	3	5	9
17			
24			
32			

Subtract 6	
36	
27	
14	
21	
18	

Draw a line from each number to the correct answer.

36 − [9 / 7 / 5] = [29 / 31 / 27]

29 + [2 / 9 / 4] = [38 / 33 / 31]

42

4D Adding two 2-digit numbers

Discover

Worked example

$$21 + 17 = 38$$

$$20 \quad 1 \quad 10 \quad 7$$

$$20 + 10 = 30 \quad 1 + 7 = 8$$

$$30 + 8 = 38$$

Split each number into tens and ones. Add the tens. Add the ones. Add the two new totals. You do not need to draw the arrows. They just show you where the numbers came from. This is the short way to write the number sentence:

$$21 + 17 = 20 + 10 + 1 + 7 = 38$$

Solve these addition number sentences.

$25 + 13 = 20 + 10 + 5 + 3 =$

$28 + 14 =$

$37 + 21 =$

$24 + 19 =$

15	61	47
74	26	82
38	53	19

+ 23 →

Complete this addition grid.

+	15	42	21	37	53	29	64
19							
33							

4D Adding two 2-digit numbers

Explore

Solve this cross-number puzzle.

Clues

Across
1. 17 + 15
2. 43 + 52
3. 28 + 13
4. 29 + 27
5. 37 + 31
7. 42 + 41
8. 58 + 39

Down
1. 25 + 12
2. 47 + 44
3. 27 + 19
4. 31 + 27
5. 38 + 26
6. 18 + 15
7. 45 + 42

Write the clues for this cross-number puzzle.

Grid values:
- Row 1: 1:4, 6, _, 2:5, 7
- Row 2: 5, _, 3:6, 1, _
- Row 3: _, 4:8, 8, _, 6:2
- Row 4: 5:9, 2, _, 7:7, 4
- Row 5: 3, _, 8:5, 9, _

Clues

Across
1. _____
2. _____
3. _____
4. _____
5. _____
7. _____
8. _____

Down
1. _____
2. _____
3. _____
4. _____
5. _____
6. _____
7. _____

The chickens laid 18 eggs on Monday and 23 eggs on Tuesday. How many eggs altogether?

There were 27 students in one class and 28 in another class. How many students altogether?

Make up a number story to go with this calculation: 25 + 19 = 44

Here is an example. There were 25 students in one class and 19 in another class. There were 44 students altogether. What will your story be about?

4E Finding the difference

Discover

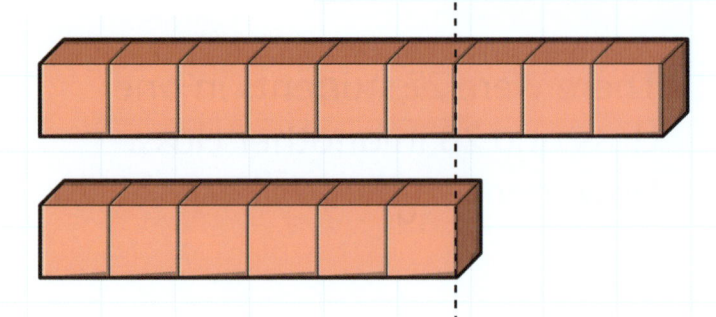

9

6

The difference between 9 and 6 is 3

9 − 6 = 3

Find the difference.

The difference between ☐ and ☐ is ☐

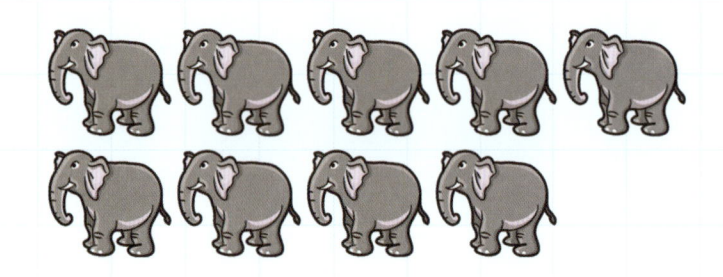

☐ − ☐ = ☐

The difference between ☐ and ☐ is ☐

☐ − ☐ = ☐

The difference between ☐ and ☐ is ☐

☐ − ☐ = ☐

The difference between 14 and 8 is ☐   14 − 8 = ☐

The difference between 19 and 7 is ☐   19 − 7 = ☐

The difference between 17 and 12 is ☐   17 − 12 = ☐

The difference between 24 and 18 is ☐   24 − 18 = ☐

The difference between 38 and 31 is ☐   38 − 31 = ☐

Make up a number story to go with this calculation: 28 − 9 = 19

Here is an example. There were 28 students in a class. 9 students were unwell so only 19 students were at school. What will your story be about?

4E Finding the difference

Explore

34 − 28 = ☐   27 − 25 = ☐

36 − 32 = ☐   22 − 18 = ☐

61 − 57 = ☐   63 − 58 = ☐

42 − 39 = ☐   31 − 28 = ☐

82 − 79 = ☐   74 − 69 = ☐

53 − 49 = ☐   45 − 38 = ☐

29 − 25 = ☐   53 − 47 = ☐

−	21	23	25
27			
28			
29			

You could **count back** from these numbers to find the difference.

You could **count on** from these numbers to find the difference.

Complete these sentences using the words **count on** or **count back**.

To find a small difference between two numbers, you can _____ from the smaller to the larger number.

To find a small difference between two numbers, you can _____ from the larger to the smaller number.

4F Missing numbers

Discover

Think: How many more to make 45?

Count up from 27 to 45, the missing number is 18.

Think: How many more to make 45?

I did it a different way.

I know 7 + 3 = 10, so 27 + 3 = 30, 30 + 10 = 40, 40 + 5 = 45.

Worked example

27 + ☐ = 45   27 + 18 = 45

45 − 27 = ☐   45 − 27 = 18

Think: Make it a subtraction

45 − 27 = ☐

45 − 20 = 25, 25 − 7 = 18

The missing number is 18.

Think: Make it a subtraction

If it was a smaller difference, I could have counted up on a number line.

Find the missing numbers.

15 + ☐ = 20   19 + ☐ = 30   ☐ + 40 = 100

☐ + 25 = 40   40 + ☐ = 57   63 + ☐ = 88

☐ + 16 = 63   ☐ + 61 = 80   ☐ + 49 = 67

Calculating – Addition and Subtraction

49

Think: Count back on a number line.

I need to count back 9 to get from 34 to 25, so the missing number is 9.

Think: How many do I need to take away to get from 34 to 25?

34 − 10 is 24, so 34 − 9 is 25.

The missing number is 9.

Worked example

34 − ☐ = 25   34 − 9 = 25

25 + ☐ = 34   25 + 9 = 34

Think: Make it an addition and count up from 25 to 34.

25 + 5 = 30, 30 + 4 = 34.

5 + 4 = 9, the missing number is 9.

Think: Count up in my head.

25 and 5 is 30, 30 and 4 is 34, so I counted on 5 and 4, that's 9.

Find the missing numbers.

25 − ☐ = 20   47 − ☐ = 30   30 − ☐ = 19

☐ − 13 = 64   ☐ − 31 = 27   ☐ − 39 = 52

86 − ☐ = 59   76 − ☐ = 34   62 − ☐ = 41

4F Missing numbers

Explore

The scales show that the totals each side of the equals sign are the same. The scales **balance**. So we could write

15 + 12 = 13 + 14

Make these scales balance

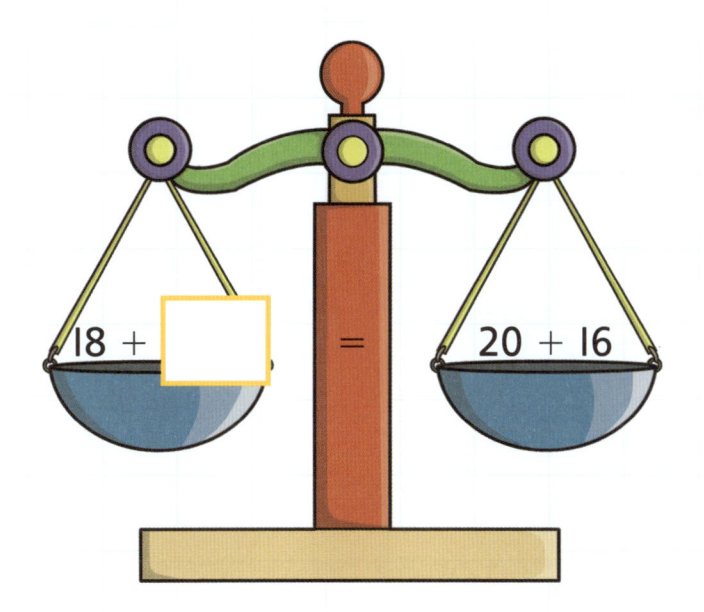

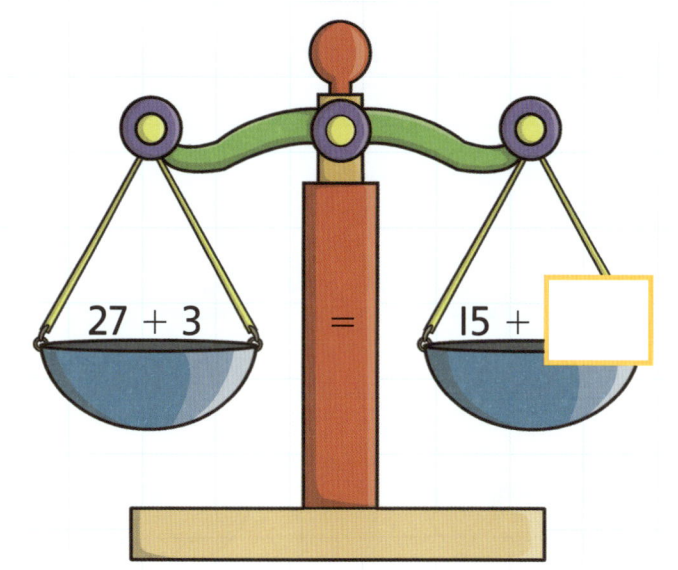

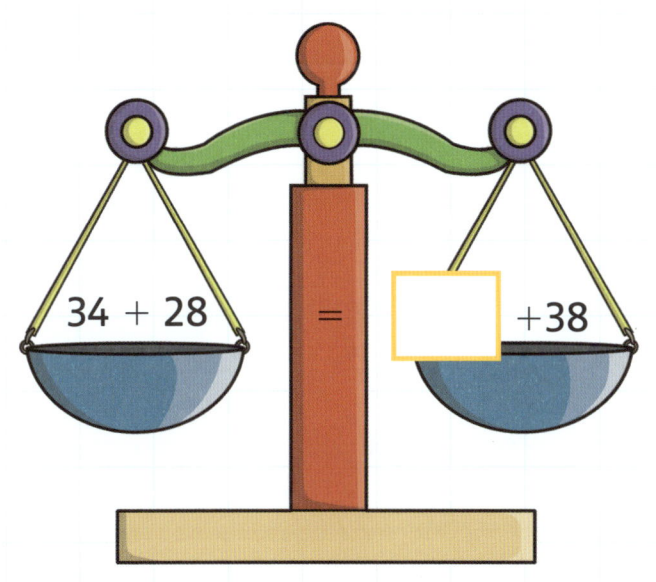

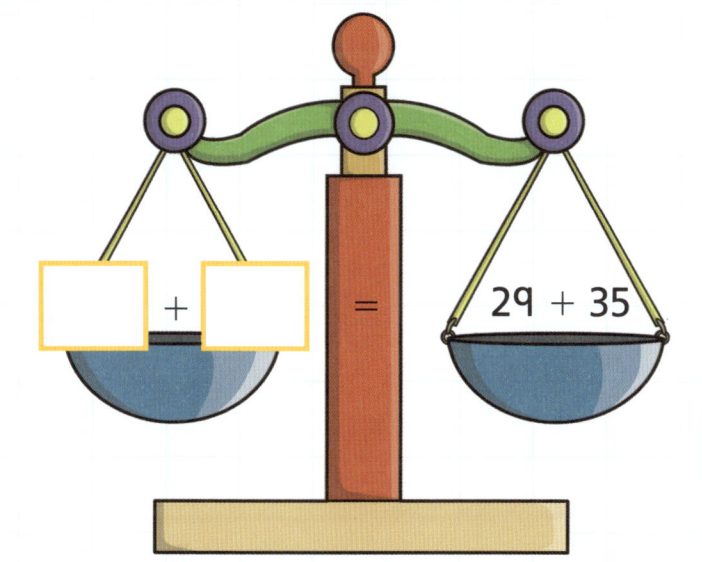

15 + 12 = 13 + 14

18 + ☐ = 20 + 16

27 + 3 = 15 + ☐

34 + 28 = ☐ + 38

☐ + ☐ = 29 + 35

26 + ☐ = 14 + ☐

17 + 26 = 15 + ☐

36 + 26 = ☐ + ☐

☐ + ☐ = ☐ + ☐

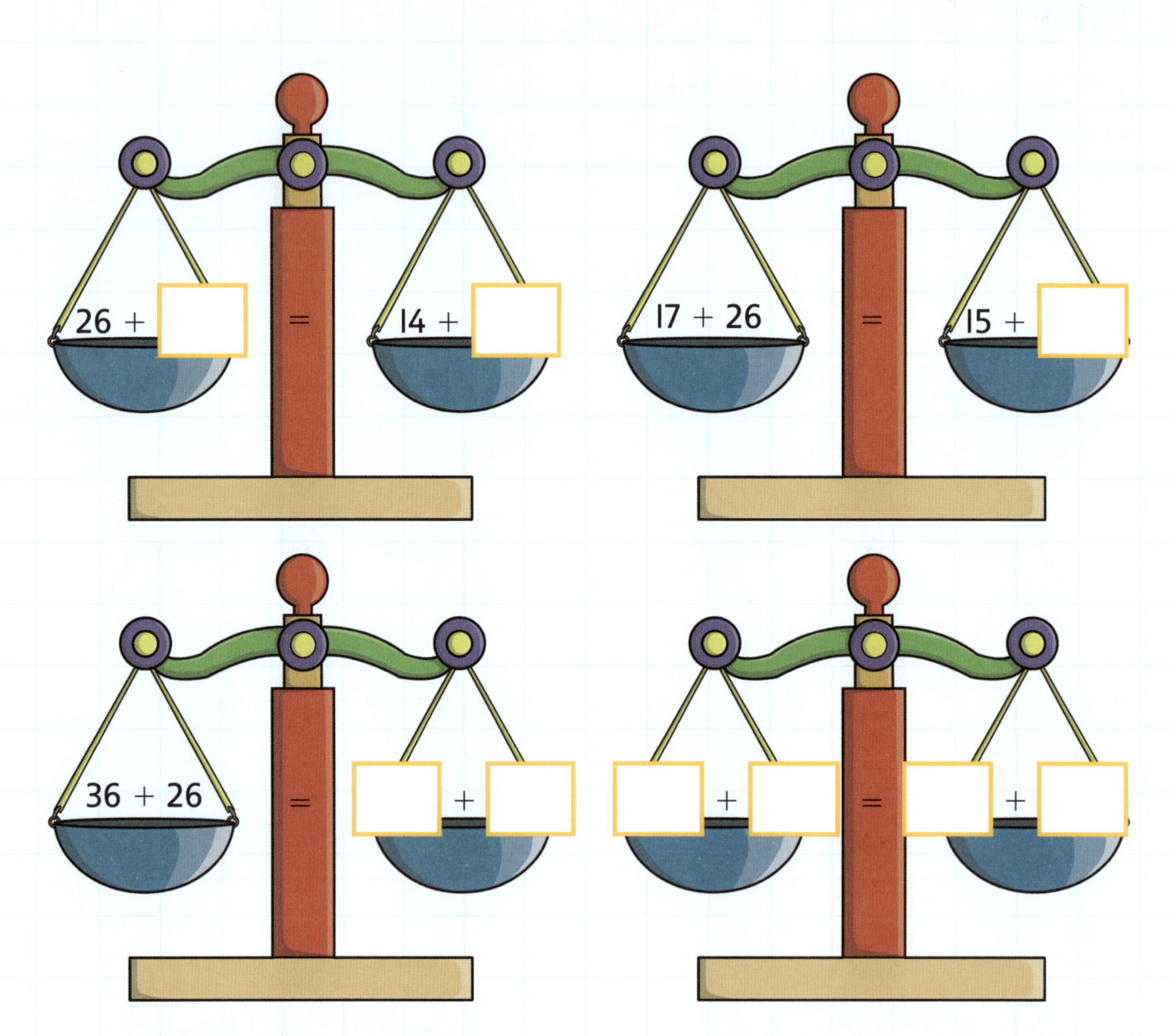

Write **true** or **false** next to each sentence.

9 + 14 = 14 + 9 _____

23 + 25 = 32 + 52 _____

68 − 31 = 23 + 14 _____

74 − 38 = 81 − 44 _____

42 + 16 = 46 + 12 _____

45 − 12 = 19 + 26 _____

26 + 24 = 25 + 25 _____

57 − 25 = 55 − 27 _____

52

4 Calculating – addition and subtraction

Connect

What happens when you take a 2-digit number, swap the digits round and **add** the two numbers?

> 32 + 23 = 30 + 20 + 2 + 3 = 55
>
> 62 + 26 = 60 + 20 + 2 + 6 = 88

Try some more numbers. What do you notice?

Colour the totals on a 100-square to help you.

When you take a 2-digit number, swap the digits round and add the two numbers, the answer _____

_____

4 Calculating – addition and subtraction

Review

What happens when you take a 2-digit number, swap the digits round and **subtract** the larger number from the smaller number?

73 − 37 = 36   82 − 28 = ☐

> Predict what you think will happen.
>
> I think _____
>
> _____
>
> _____

Try some numbers. What do you notice?

When you take a 2-digit number, swap the digits round and subtract the smallest number from the largest number, the answer

_____

_____

5 Number Families

Engage

What is a fact family?

There are lots of different people in my family, so there must be more than two facts in a **fact family**.

But everyone in your family is related to each other, so the number facts in a **fact family** must be related too.

But *how* are they related?

5A Fact families for number pairs to 20

Discover

First house (purple):
- 20
- 1 + 19
- 19 + 1 = 20
- 1 + 19 = 20
- 20 − 19 = 1
- 20 − 1 = 19

Which fact family did not need all the floors in the house? _____

5A Fact families for number pairs to 20

Explore

Complete these sentences.

I found the fact families for number _____.

There are _____ different families.

Write any fact family which did not need all the floors in the house here.

_____

5B Fact families for multiples of 10 to 100

Discover

Number pairs for 10	Multiply each number by 10 to make it 10 times bigger, so each number pair for 10 becomes a number pair for 100 using multiples of 10	Number pairs for 100 using multiples of 10
10 + 0 = 10		100 + 0 = 100

Write the two related subtraction facts for this number pair:

80 + 20 = 100, 20 + 80 = 100 _____

Circle the number which does not belong to the fact family.
The first one has been done for you.

100	20	(70)	80

6	20	10	4

4	2	8	10

100	50	100	50

40	60	30	100

50	100	0	100

0	10	10	1

9	10	1	3

10	7	4	3

70	100	40	30

10	40	100	90

5	10	5	0

Complete the following sentence using the word **bigger** or **smaller**.

Number pairs for 100 using multiples of 10 are the same as number pairs for 10, they are just ten times

_____ .

58

5B Fact families for multiples of 10 to 100

Explore

Write the fact family for each number pair for 100 using multiples of 10.

Which fact family did not need all the floors in the house?

_____.

Number Families

5 Number families

Connect

Look at these statements about fact families. Write **true** or **false** next to each statement and give at least three examples to show that your answer is correct. The first one has been done for you.

All fact families have four related facts.   false

①   ②   ③

5 + 5 = 10   10 + 10 = 20   3 + 3 = 6

10 − 5 = 5   20 − 10 = 10   6 − 3 = 3

There are only two facts in these fact families.

Each set of number pairs has a fact family with only two facts.

Each fact family can be recorded as a triangle.

5 Number families

Review

Here is an alien number sentence.

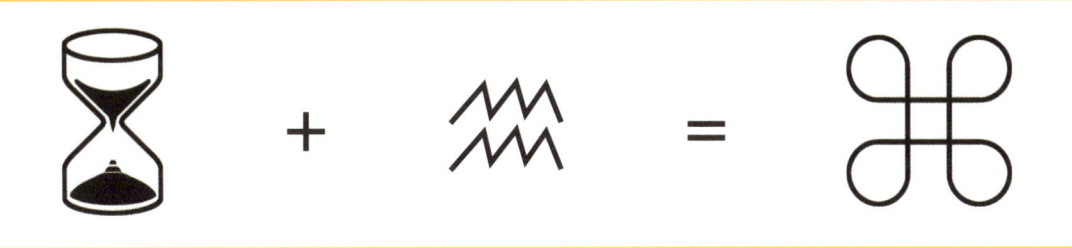

⏳ + 〰 = ⌘

They use add (+), subtract (−) and equals (=) just like we do.

Write the other three number sentences in the alien fact family.

6 Multiplication and Division

Engage

Special offer: Trainers $30 a pair

How many?

Special offer: Coloured pencils $2 pack of 10

10 packs of 10

10 packs of 10

10 packs of 10

Special offer: 5 cakes $1 pack of 5

Special offer: 5 cakes $1 pack of 5

6A Twos

Discover

0	1	2	3	4	5	6	7	8	9	10
	11	12	13	14	15	16	17	18	19	20
	21	22	23	24	25	26	27	28	29	30
	31	32	33	34	35	36	37	38	39	40
	41	42	43	44	45	46	47	48	49	50
	51	52	53	54	55	56	57	58	59	60
	61	62	63	64	65	66	67	68	69	70
	71	72	73	74	75	76	77	78	79	80
	81	82	83	84	85	86	87	88	89	90
	91	92	93	94	95	96	97	98	99	100

2

Complete the sentence using the word **even** or **odd**.

When you count in twos from 0, all the numbers are _____.

Start at 0. Draw 3 jumps of 2.

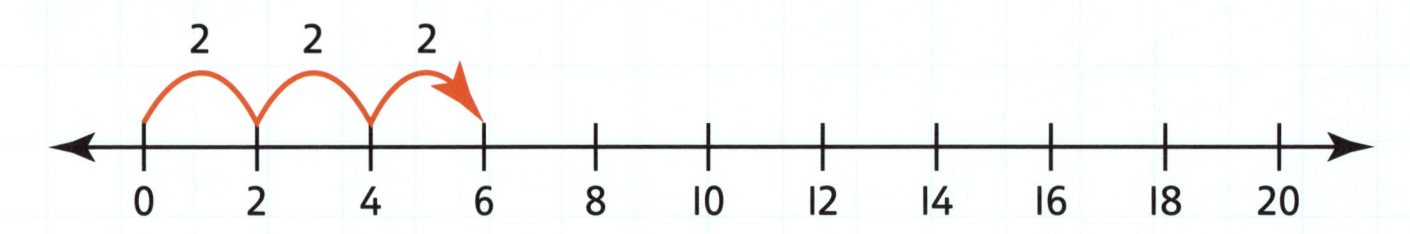

Which numbers do you land on? 2, 4, 6

Which number did you finish on? 6

64

Start at 0. Draw 5 jumps of 2.

Which numbers do you land on? ☐

Which number did you finish on? ☐

```
←—+—+—+—+—+—+—+—+—+—+—+→
  0  2  4  6  8  10 12 14 16 18 20
```

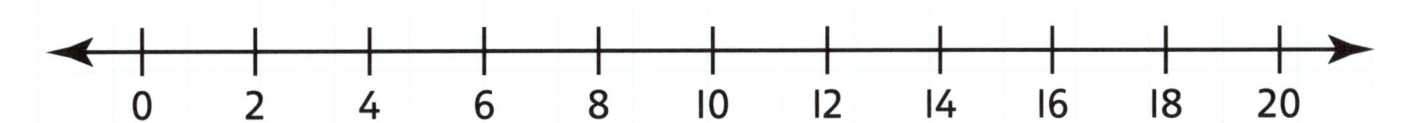

Start at 0. Draw 8 jumps of 2.

Which numbers do you land on? ☐

Which number did you finish on? ☐

```
←—+—+—+—+—+—+—+—+—+—+—+→
  0  2  4  6  8  10 12 14 16 18 20
```

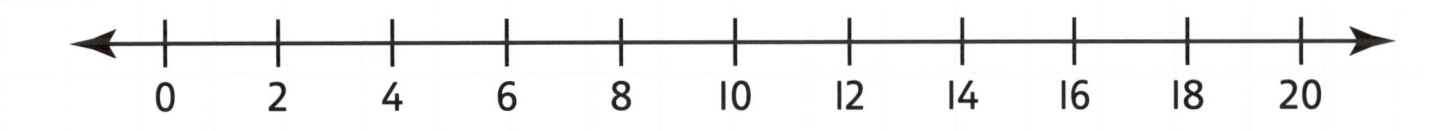

6A Twos

Explore

Complete the two times table up to 10 × 2.

0 × 2 = 0	
1 × 2 = 2	2
2 × 2 = 4	2 + 2 = 4
3	2 + 2 + 2 =
4	
5	
6	
7	
8	
9	
10	

Complete this sentence using either **odd** or **even**.

All the totals in the two times table are _____.

6B Fives and tens

Discover

0	1	2	3	4	5	6	7	8	9	10
	11	12	13	14	15	16	17	18	19	20
	21	22	23	24	25	26	27	28	29	30
	31	32	33	34	35	36	37	38	39	40
	41	42	43	44	45	46	47	48	49	50
	51	52	53	54	55	56	57	58	59	60
	61	62	63	64	65	66	67	68	69	70
	71	72	73	74	75	76	77	78	79	80
	81	82	83	84	85	86	87	88	89	90
	91	92	93	94	95	96	97	98	99	100

5

Complete the sentence using two of the words **zero**, **one**, **two** and **five**.

When you count in fives from 0, all the ones have _____ or _____.

0	1	2	3	4	5	6	7	8	9	10
	11	12	13	14	15	16	17	18	19	20
	21	22	23	24	25	26	27	28	29	30
	31	32	33	34	35	36	37	38	39	40
	41	42	43	44	45	46	47	48	49	50
	51	52	53	54	55	56	57	58	59	60
	61	62	63	64	65	66	67	68	69	70
	71	72	73	74	75	76	77	78	79	80
	81	82	83	84	85	86	87	88	89	90
	91	92	93	94	95	96	97	98	99	100

10

Complete the sentence using either **zero** or **five**.

When you count in tens from 0, all the ones have _____ or _____.

Multiplication and Division

Start at 0. Draw 4 jumps of 5.

Which numbers do you land on? ⬜

Which number did you finish on? ⬜

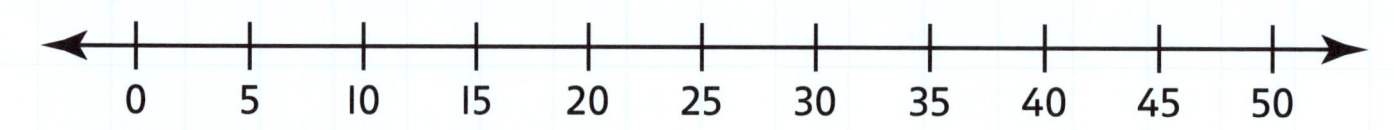

←—+——+——+——+——+——+——+——+——+——+——+→
　0　　5　　10　　15　　20　　25　　30　　35　　40　　45　　50

Start at 0. Draw 4 jumps of 10.

Which numbers do you land on? ⬜

Which number did you finish on? ⬜

←—+——+——+——+——+——+——+——+——+——+——+→
　0　　10　　20　　30　　40　　50　　60　　70　　80　　90　　100

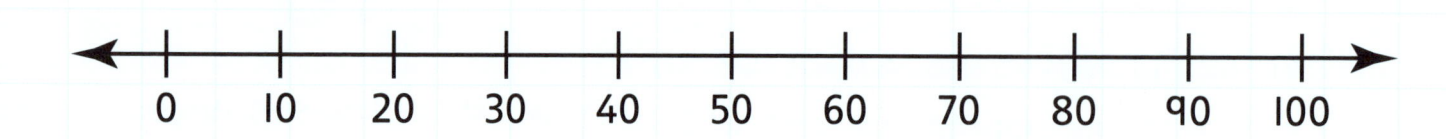

6B Fives and tens

Explore

Complete the five times table up to 10 × 5.

0 × 5 = 0	
1 × 5 = 5	5
2 × 5 = 10	5 + 5 = 10
3	5 + 5 + 5 =
4	
5	
6	
7	
8	
9	
10	

Complete this sentence using two of the words **zero**, **one**, **two** and **five**.

When you count in fives from 0, all the ones have _____ or _____.

Complete the ten times table up to 10 × 10.

0 × 10 = 0	
1 × 10 = 10	10
2 × 10 = 20	10 + 10 = 20
3	10 + 10 + 10 =
4	
5	
6	
7	
8	
9	
10	

Complete this sentence using either **zero** or **five**.

When you count in tens from 0, all the ones have _____ or _____.

6C Threes and fours

Discover

0	1	2	3	4	5	6	7	8	9	10
	11	12	13	14	15	16	17	18	19	20
	21	22	23	24	25	26	27	28	29	30
	31	32	33	34	35	36	37	38	39	40
	41	42	43	44	45	46	47	48	49	50
	51	52	53	54	55	56	57	58	59	60
	61	62	63	64	65	66	67	68	69	70
	71	72	73	74	75	76	77	78	79	80
	81	82	83	84	85	86	87	88	89	90
	91	92	93	94	95	96	97	98	99	100

3

Complete the sentence using the word **straight** or **diagonal**.

When you count in threes, the numbers make _____ lines on the 100-square.

0	1	2	3	4	5	6	7	8	9	10
	11	12	13	14	15	16	17	18	19	20
	21	22	23	24	25	26	27	28	29	30
	31	32	33	34	35	36	37	38	39	40
	41	42	43	44	45	46	47	48	49	50
	51	52	53	54	55	56	57	58	59	60
	61	62	63	64	65	66	67	68	69	70
	71	72	73	74	75	76	77	78	79	80
	81	82	83	84	85	86	87	88	89	90
	91	92	93	94	95	96	97	98	99	100

4

Complete the sentence using the word **even** or **odd**.

When you count in fours from 0, all the number are _____ .

Multiplication and Division

Start at 0. Draw 4 jumps of 3.

Which numbers do you land on? ☐

Which number did you finish on? ☐

```
←——+——+——+——+——+——+——+——+——+——+——+→
   0  3  6  9 12 15 18 21 24 27 30
```

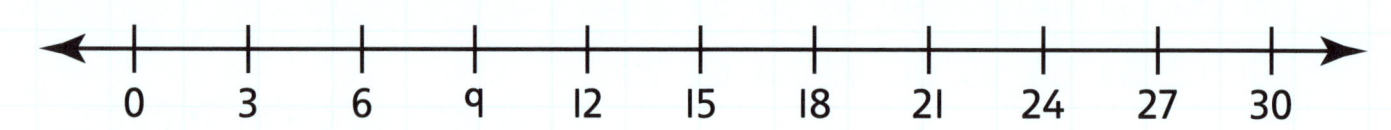

Start at 0. Draw 6 jumps of 3.

Which numbers do you land on? ☐

Which number did you finish on? ☐

```
←——+——+——+——+——+——+——+——+——+——+——+→
   0  3  6  9 12 15 18 21 24 27 30
```

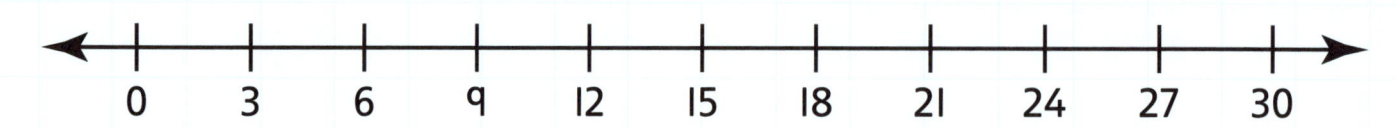

Start at 0. Draw 7 jumps of 4.

Which numbers do you land on? ☐

Which number did you finish on? ☐

```
←——+——+——+——+——+——+——+——+——+——+——+→
   0  4  8 12 16 20 24 28 32 36 40
```

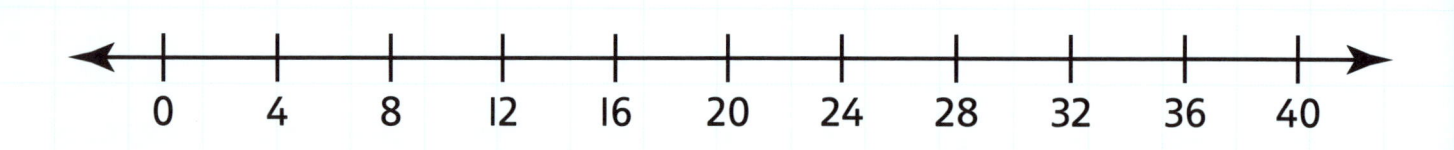

6C Threes and fours

Explore

Complete the three times table up to 10 × 3.

0 × 3 = 0	
1 × 3 = 3	3
2 × 3 = 6	3 + 3 = 6
	3 + 3 + 3 =

Complete the four times table up to 10 × 4.

0 × 4 = 0	
1 × 4 = 4	4
2 × 4 = 8	4 + 4 = 8
	4 + 4 + 4 =

6D Arrays

Discover

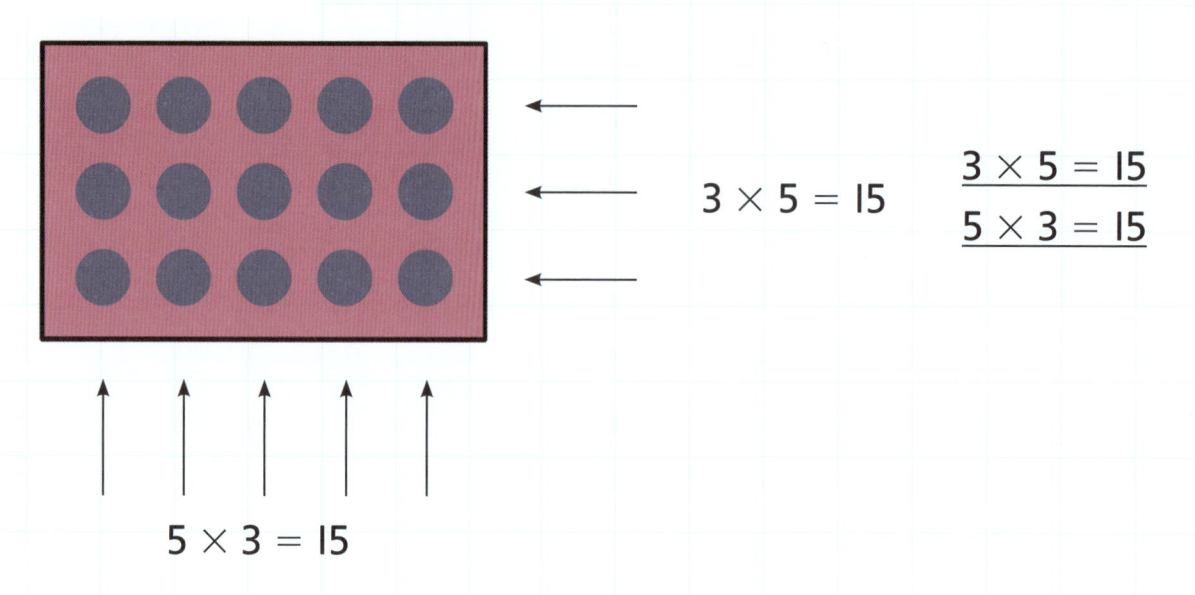

$3 \times 5 = 15$

$5 \times 3 = 15$

$\dfrac{3 \times 5 = 15}{5 \times 3 = 15}$

Write two multiplications for each array.

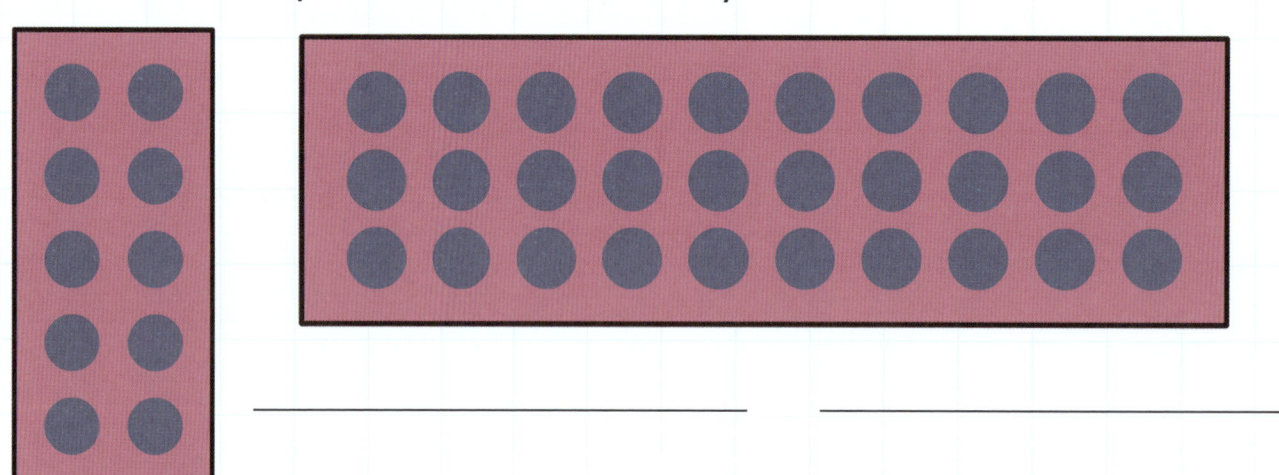

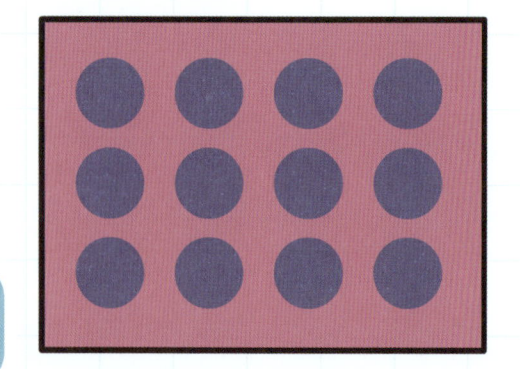

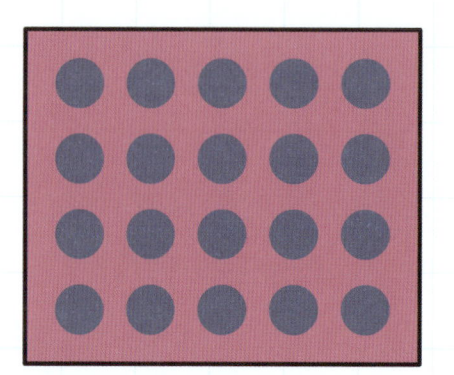

Multiplication and Division

75

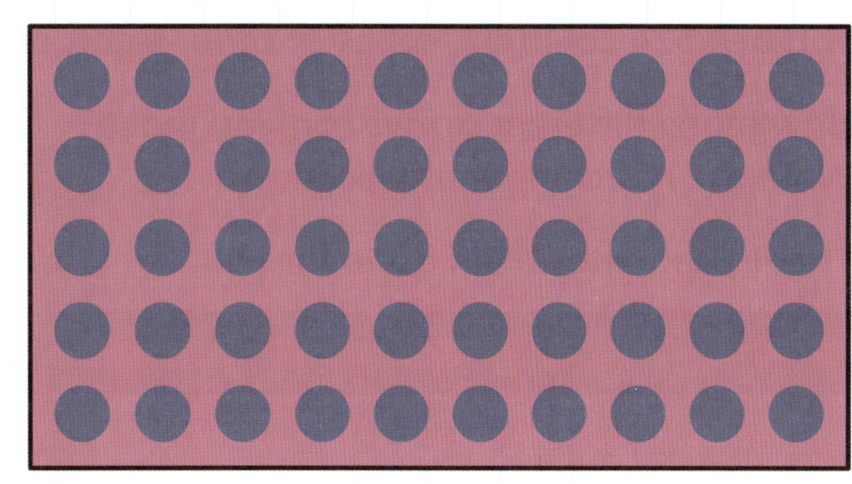

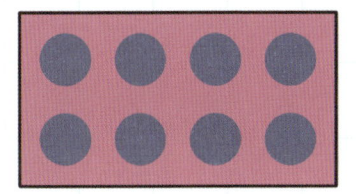

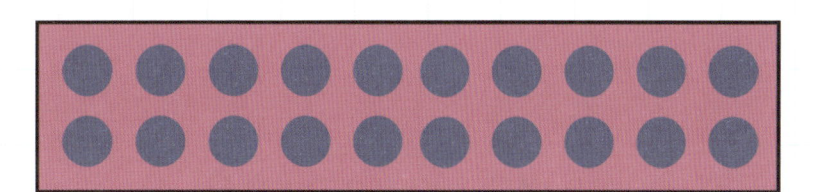

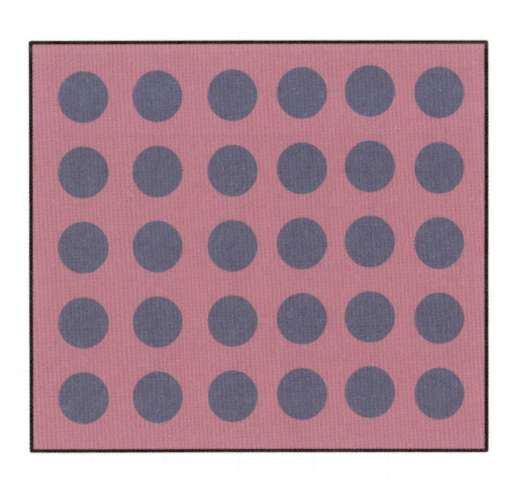

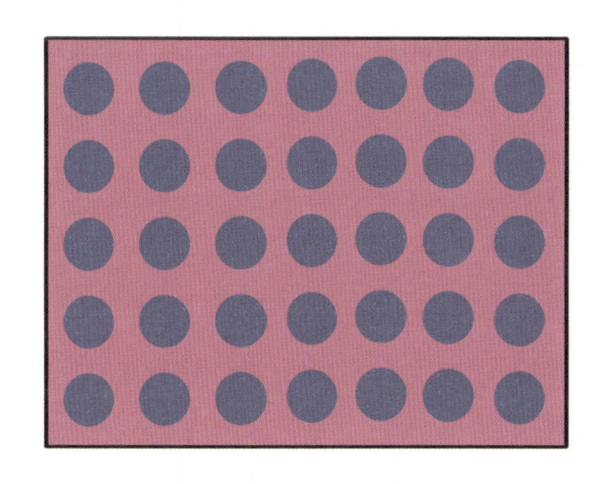

24 counters put into groups of 8.

24 ÷ 8 = _____

24 counters put into groups of 3.

24 ÷ 3 = _____

15 counters put into groups of 3.

15 ÷ 3 = _____

15 counters put into groups of 5.

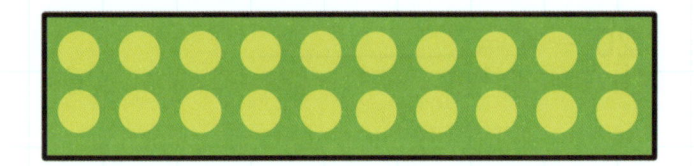

15 ÷ 5 = _____

20 counters put into groups of 10.

20 ÷ 10 =

20 counters put into groups of 2.

20 ÷ 2 =

6E Division as grouping

Explore

Find the matching divisions for each multiplication in the 2, 5 and 10 times tables. Draw the array to help you.

0 × 2 = 0	2 × 0 = 0	\multicolumn{2}{l}{Try dividing by zero on a calculator. What happens?}
1 × 2 = 2	2 × 1 = 2	2 ÷ 1 = 2	2 ÷ 2 = 1
2 × 2 = 4			
3			
4			
5			
6			
7			
8			
9			
10			

0 × 5 = 0	5 × 0 = 0	\multicolumn{2}{l}{Try dividing by zero on a calculator. What happens?}
1 × 5 = 5	5 × 1 = 5	5 ÷ 1 = 5	5 ÷ 5 = 1
2 × 5 = 10			
3			
4			
5			
6			
7			
8			
9			
10			

Multiplication and Division

0 × 10 = 0	10 × 0 = 0	Try dividing by zero on a calculator. What happens?	
1 × 10 = 10	10 × 1 = 10	10 ÷ 1 = 10	10 ÷ 10 = 1
2 × 10 = 20			
3			
4			
5			
6			
7			
8			
9			
10			

Talk to your partner (or group) about the patterns you have noticed. Describe one of the patterns.

6F Remainders

Discover

Share 10 biscuits between 3 students.

3 each and 1 left over

10 ÷ 3 = 3 r 1

Draw the groups and write the calculation for these problems. Remember to show any remainder.

Share 18 sweets between 4 students.

Put 21 balls into groups of 5.

Put 18 plates into piles of 5.

Put 14 students into groups of 3.

82

6F Remainders

Explore

Take a large handful of small objects.
Can you put the objects into groups of 2, 3, 4, 5 and 10?

Complete the sentences to show what you found out.

_____ objects put into groups of 2. There are _____ groups and _____ left over.

_____ ÷ 2 = _____

_____ objects put into groups of 3. There are _____ groups and _____ left over.

_____ ÷ 3 = _____

_____ objects put into groups of 4. There are _____ groups and _____ left over.

_____ ÷ 4 = _____

_____ objects put into groups of 5. There are _____ groups and _____ left over.

_____ ÷ 5 = _____

_____ objects put into groups of 10. There are _____ groups and _____ left over.

_____ ÷ 10 = _____

Now choose groups of a different number to try.
Complete the sentences to show what you did.

_____ objects put into groups of _____. There are _____ groups and _____ left over.

_____ ÷ _____ = _____

_____ objects put into groups of _____. There are _____ groups and _____ left over.

_____ ÷ _____ = _____

_____ objects put into groups of _____. There are _____ groups and _____ left over.

_____ ÷ _____ = _____

_____ objects put into groups of _____. There are _____ groups and _____ left over.

_____ ÷ _____ = _____

_____ objects put into groups of _____. There are _____ groups and _____ left over.

_____ ÷ _____ = _____

6 Multiplication and division

Connect

How many students are in your class today?

The teacher is going to put all the students in the class into groups of 2. Is there anyone without a partner?

The next activities need groups of 3, then 4, then 5, then 10. Will there be anyone who is not in a group?

Draw a picture to show your groups of 2, then 3, 4, 5 and 10. Write the matching division calculation next to each picture.

6 Multiplication and division

Review

Write the matching times table, up to 10 times, below each of these three numbers.

2

5

10

3

Count in threes. Label each mark on the number line.

0

4

Count in fours. Label each mark on the number line.

0

Write the fact family for this triangle.

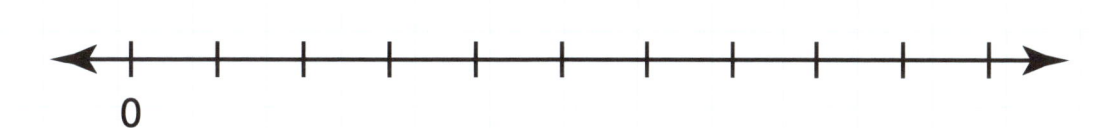

2, ×, ÷, 5, 10

_____

_____

Write the fact family for this triangle.

8, ×, ÷, 5, 40

_____

_____

There are 15 biscuits in a packet. Four students share the biscuits. They each get the same number.

How many biscuits do they get each? _____

How many biscuits are left over? _____

Write this in a number sentence. _____

_____

Write a number story for this calculation: 14 ÷ 3 = 4 r2

_____

_____

_____

7 Parts of a Whole

Engage

That's not fair!

But you've got more than me!

Here's yours, Mum said you could have half.

Are you saying that's not a half?

No, but it's not half of everything!

7A Half of a shape

Discover

Colour half of each shape. Find a different way to show half of each shape. Label each half with the word **half** or the fraction $\frac{1}{2}$.

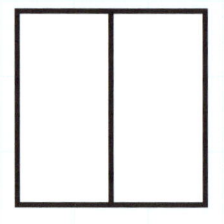

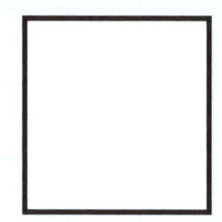

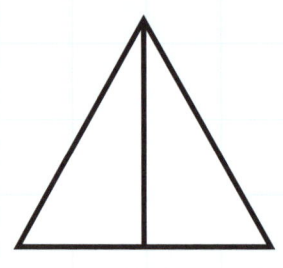

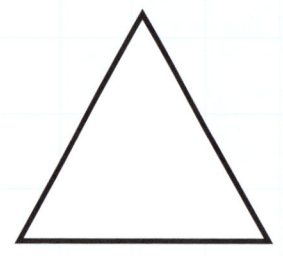

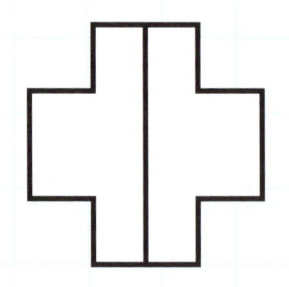

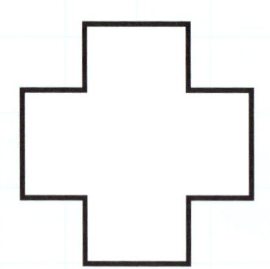

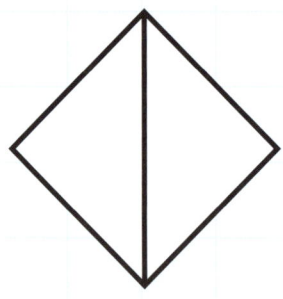

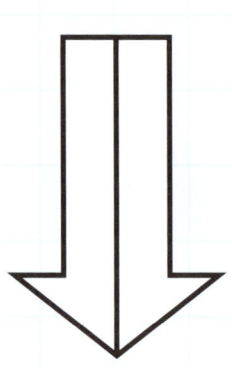

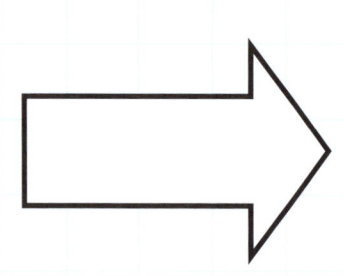

half

one out of two equal pieces

$\frac{1}{2}$

90

7A Half of a shape

Explore

Find half of each shape in two different ways.
Label each half with the word **half** or the fraction $\frac{1}{2}$.

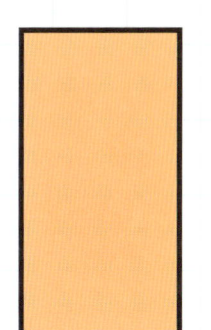

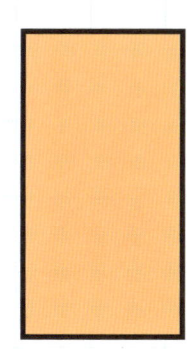

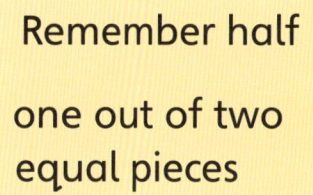

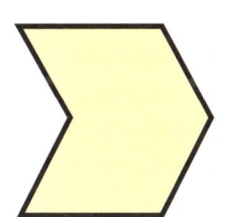

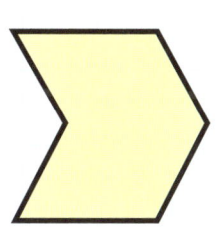

Remember half one out of two equal pieces $\frac{1}{2}$

Tick ✓ the shapes that show half.

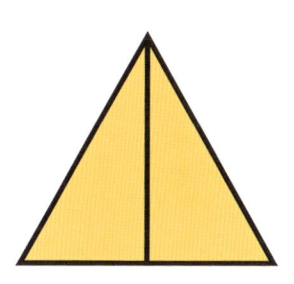

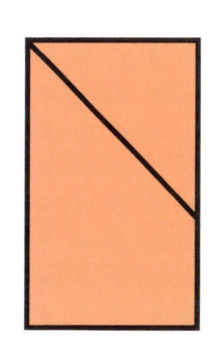

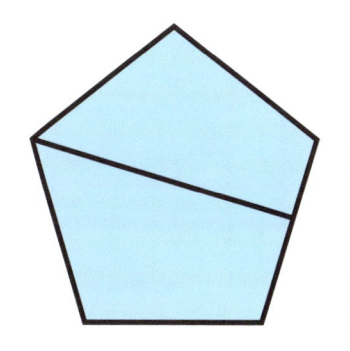

7B Half of an amount

Discover

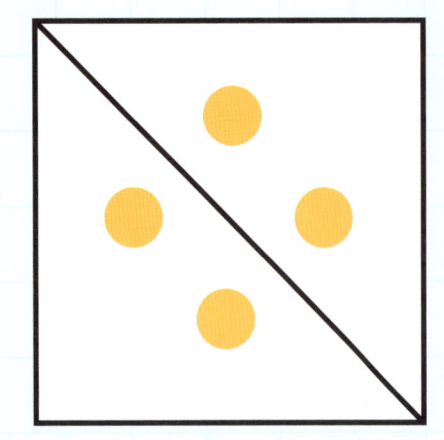

Half of 4 is 2

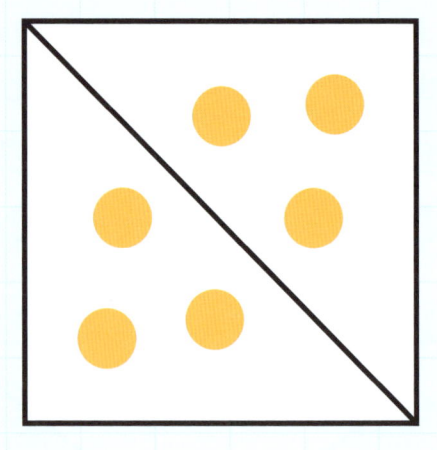

Half of _____ is _____

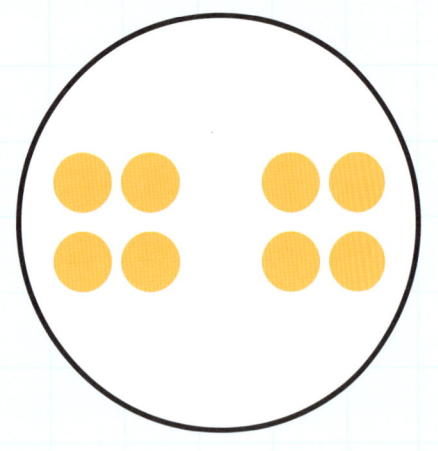

Half of _____ is _____

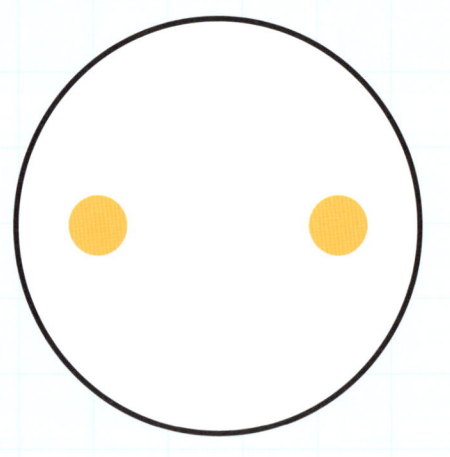

Half of _____ is _____

Complete this table.

number	half
2	
	2
6	
	4
10	

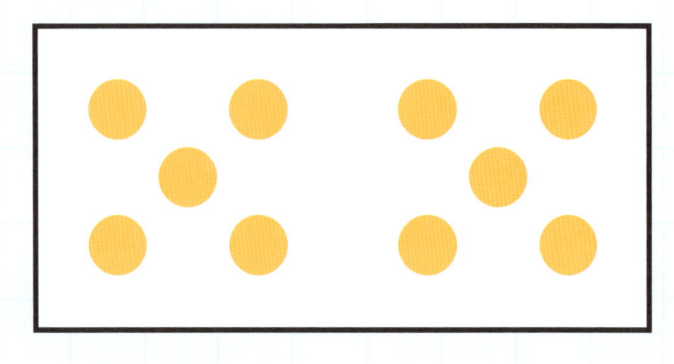

Half of _____ is _____

7A Half of an amount

Explore

Count the sweets, then draw a circle around half of them.
The first one is done for you.

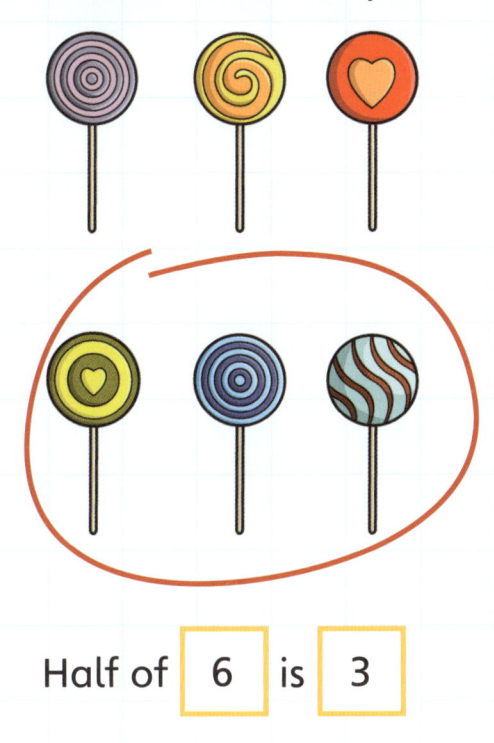

Half of ⬚6⬚ is ⬚3⬚

$\frac{1}{2}$ of ⬚ is ⬚

$\frac{1}{2}$ of ⬚ is ⬚

Half of ⬚ is ⬚

Parts of a Whole

93

Half of ☐ is ☐

½ of ☐ is ☐

½ of ☐ is ☐

Half of ☐ is ☐

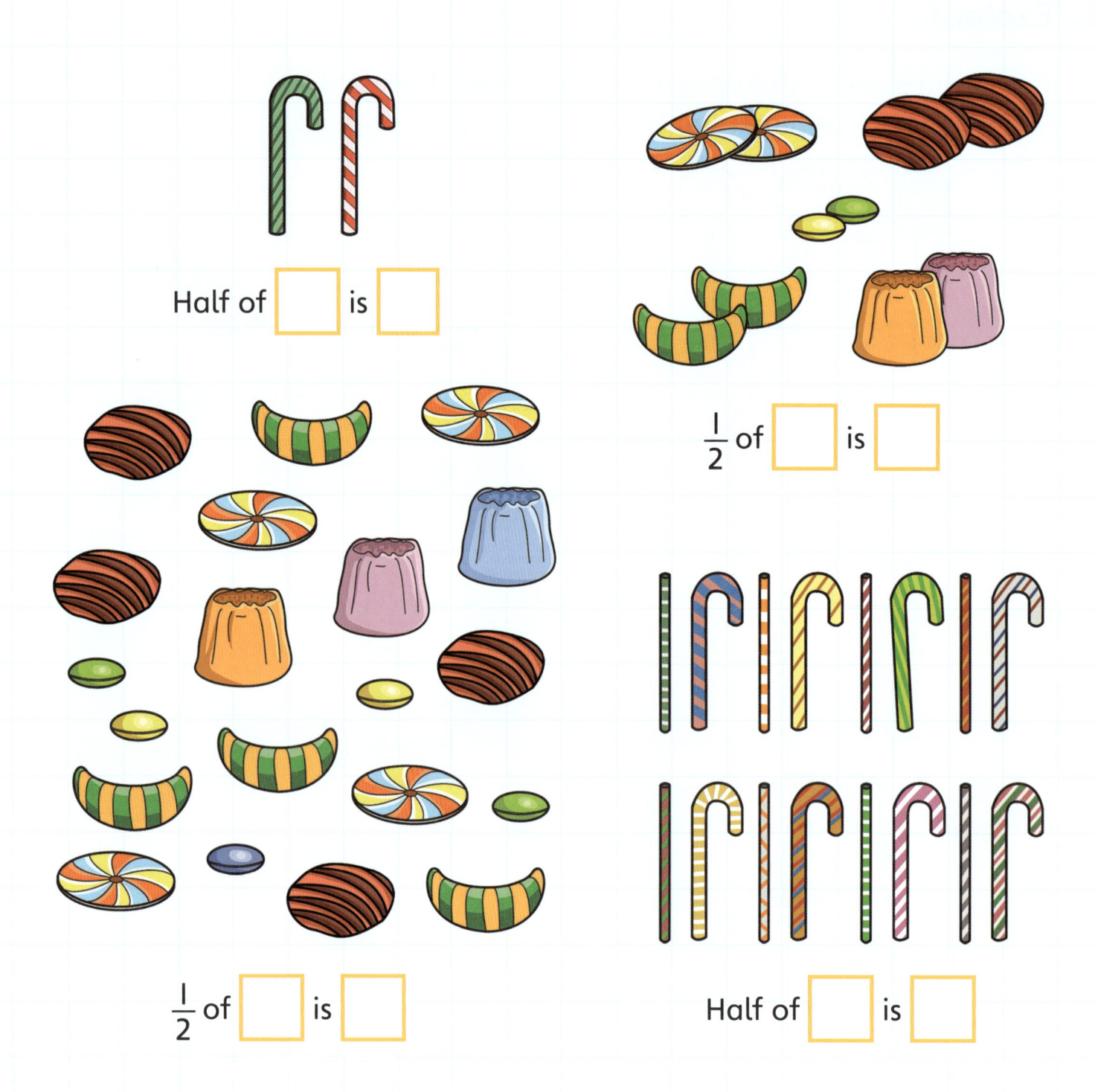

Use some counters for sweets. Find half of 18.

½ of 18 = ☐

7C Quarter and three-quarters of a shape

Discover

Colour one quarter of each shape. Find a different way to show a quarter of each shape. Label one quarter of each piece with the word **quarter** or the fraction $\frac{1}{4}$.

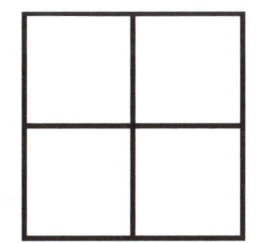

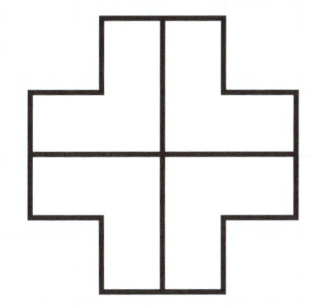

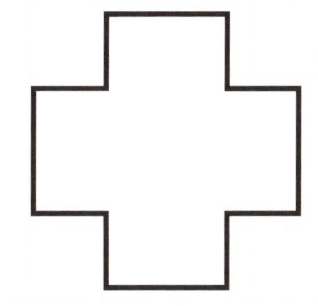

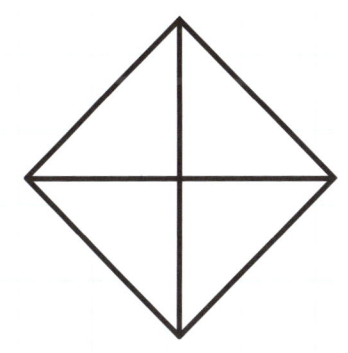

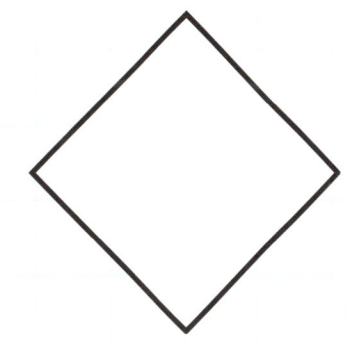

quarter
one out of four equal pieces
$\frac{1}{4}$

Colour $\frac{1}{4}$ of the shape.

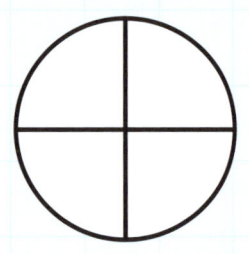

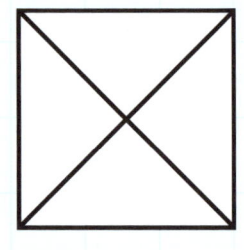

Colour $\frac{3}{4}$ of the shape.

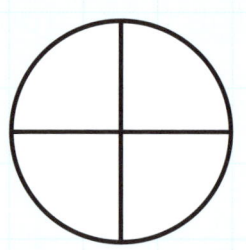

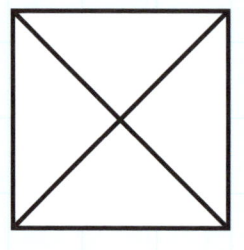

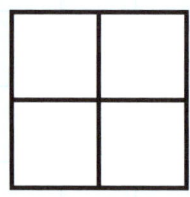

three- quarters

three out of four equal pieces

$\frac{3}{4}$

Tick ✓ the shapes that show quarters.

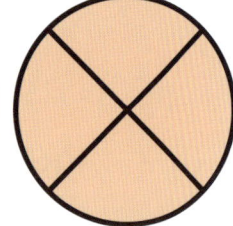

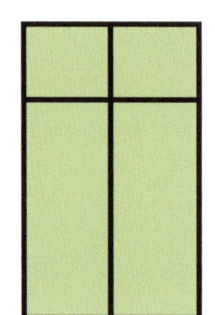

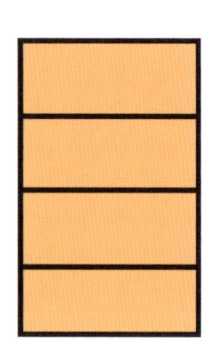

7C Quarter and three-quarters of a shape

Explore

Fold paper shapes in half and half again to make quarters.
Tick ✓ the new shapes you can make.

shape	I can make		
	square	rectangle	triangle
square			
rectangle			

Fold paper copies of these shapes into quarters.
Draw or glue in one of the quarters.

shape	quarter
circle	
hexagon	
octagon	

Complete these sentences using the words **one**, **two** or **four**.

_____ halves make _____ whole.

_____ quarters make _____ whole.

_____ half and _____ quarters are the same.

7D Quarter of an amount

Discover

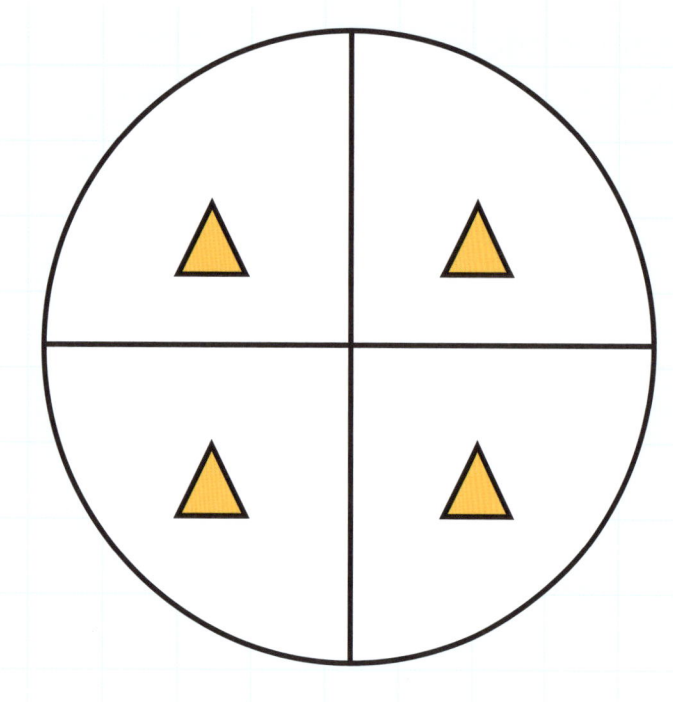

A quarter of 4 is 1

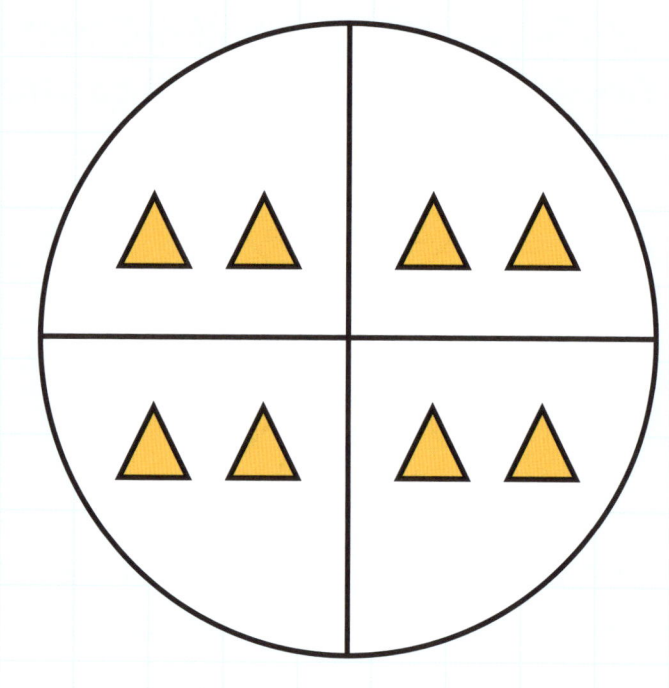

A quarter of _____ is _____

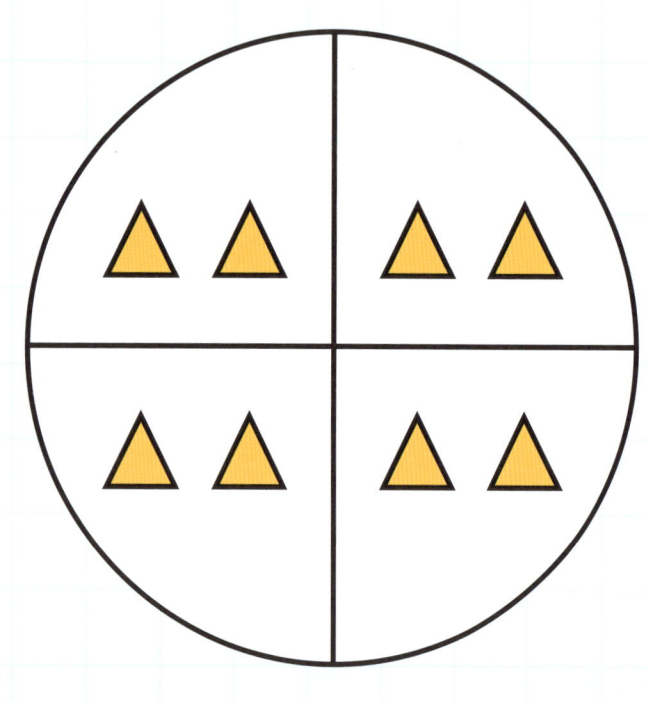

A quarter of _____ is _____

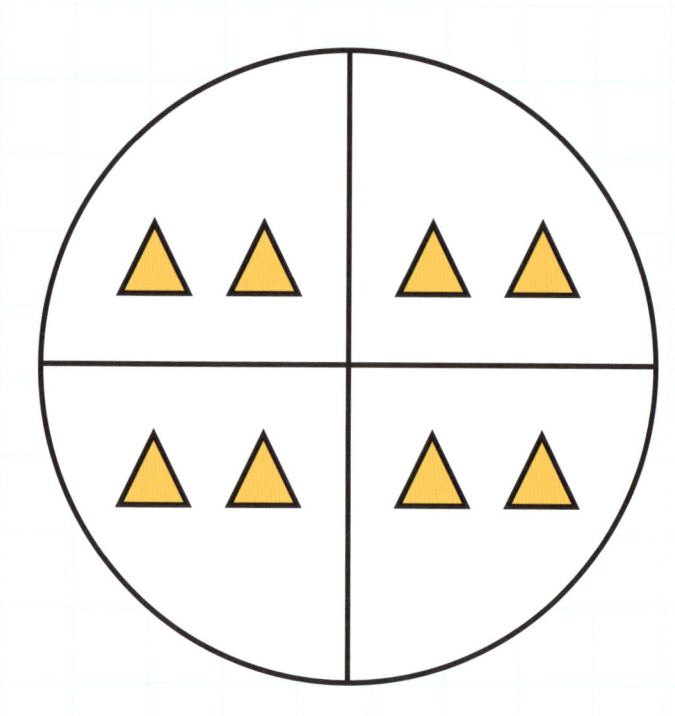

A quarter of _____ is _____

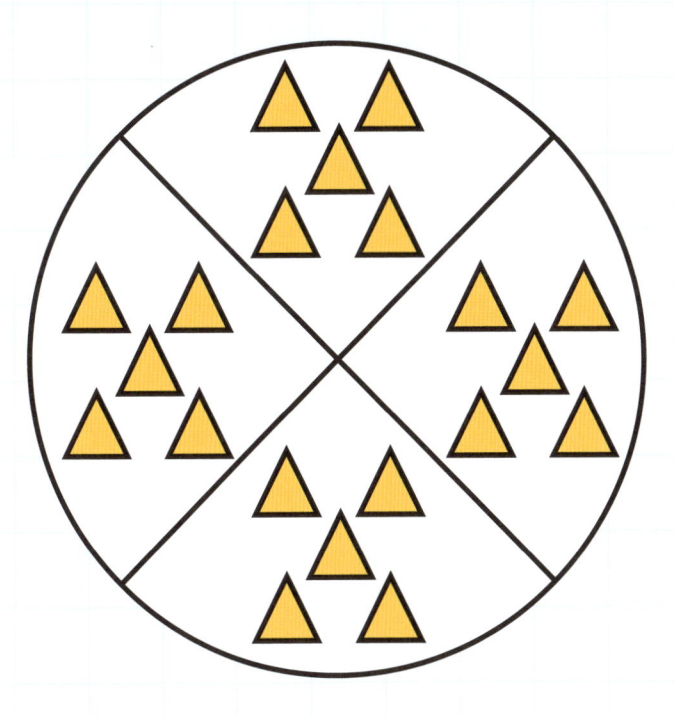

A quarter of _____ is _____

Complete this table.

number	quarter
4	1
8	
12	
	4
20	

7D Quarter of an amount

Explore

Count each set of pictures, then draw a circle around a quarter of them.

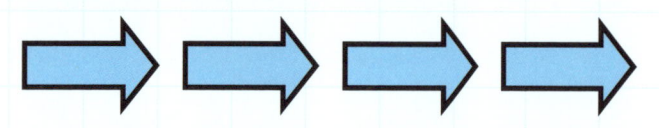

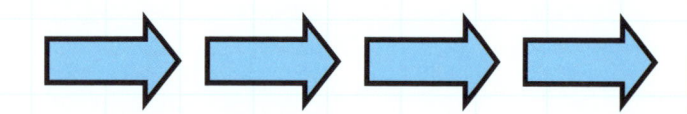

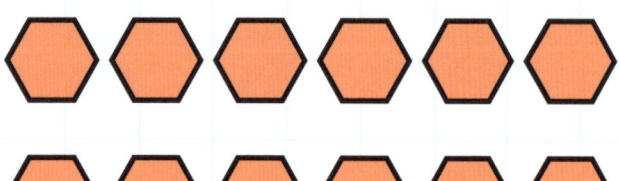

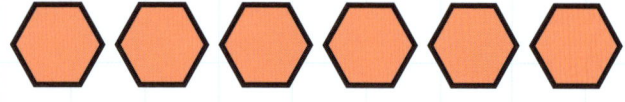

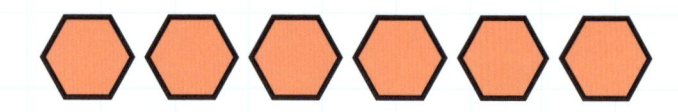

A quarter of ☐ is ☐

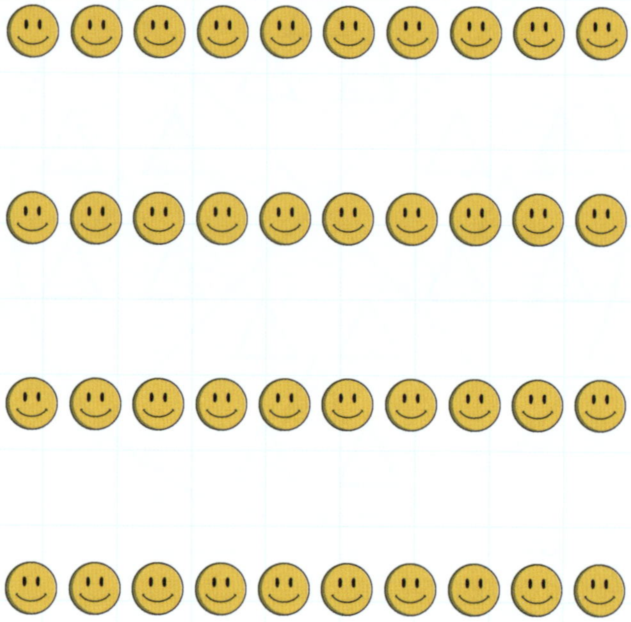

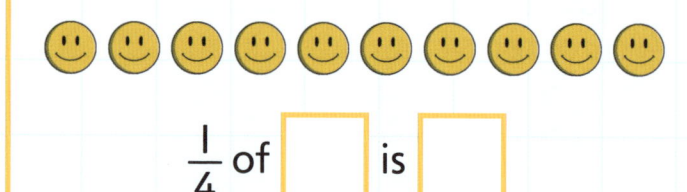

$\frac{1}{4}$ of ☐ is ☐

$\frac{1}{4}$ of ☐ is ☐

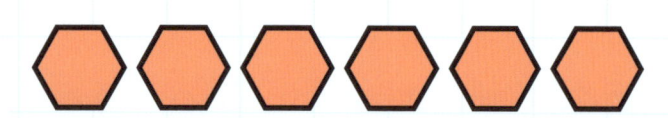

A quarter of ☐ is ☐

Use counters or cubes to help you find a quarter of 36, 28 and 32.

A quarter of 36 is _____

$\frac{1}{4}$ of 28 is _____

One quarter of 32 is _____

Complete these sentences using the words **two** or **four**.

A quarter is one of _____ equal amounts.

_____ quarters make the whole amount.

_____ quarters are the same as a half.

Parts of a Whole

101

7 Parts of a whole

Connect

12 sandwiches and 12 cakes were made for a party. Each sandwich and cake has been cut into quarters.

Everyone has a quarter of a sandwich and a quarter of a cake. How many people were at the party?

Make some notes or drawings to help you work it out.

This time everyone has half a sandwich and half a cake. How many people were at the party?

Make some notes or drawings to help you work it out.

7 Parts of a whole

Review

What if everyone was hungry and had three-quarters of a sandwich and three-quarters of a cake? How many people were at the party?

Use some counters or cubes to help you complete these number sentences.

Half of 10 is ☐   A quarter of 12 is ☐

$\frac{1}{4}$ of 16 = ☐ $\frac{1}{2}$ of 18 = ☐

Complete the sentences using **two** or **four**.

_____ halves make a whole.

_____ quarters make a whole.

_____ quarters are the same as a half.

Fold some paper shapes into halves or quarters. Cut out and glue these pieces onto this page. Label your pieces, for example, quarter of a circle, half of a hexagon, $\frac{3}{4}$ of a rectangle.

$\frac{1}{2}$ of a circle $\frac{1}{4}$ of a square

8 Shapes Everywhere

Engage

This shopping reminds me of the solid shapes at school.

This tin has got two round ends but I can't remember what the shape is called.

This one looks like a pyramid. There's one of those in our shapes.

The melon looks like a ball. What is the mathematical name for it?

105

8A 2D shapes

Discover

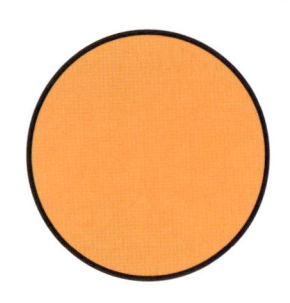

This is a _____.
It has _____ sides and _____ vertices.

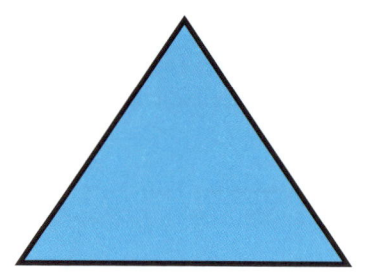

This is a _____.
It has _____ sides and _____ vertices.

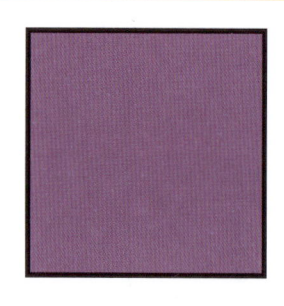

This is a _____.
It has _____ sides and _____ vertices.

This is a _____.
It has _____ sides and _____ vertices.

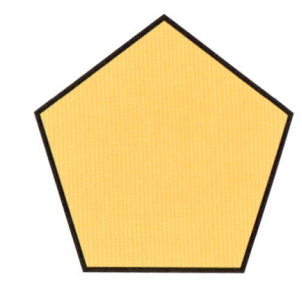

This is a _____.
It has _____ sides and _____ vertices.

This is a _____.
It has _____ sides and _____ vertices.

8A 2D shapes

Explore

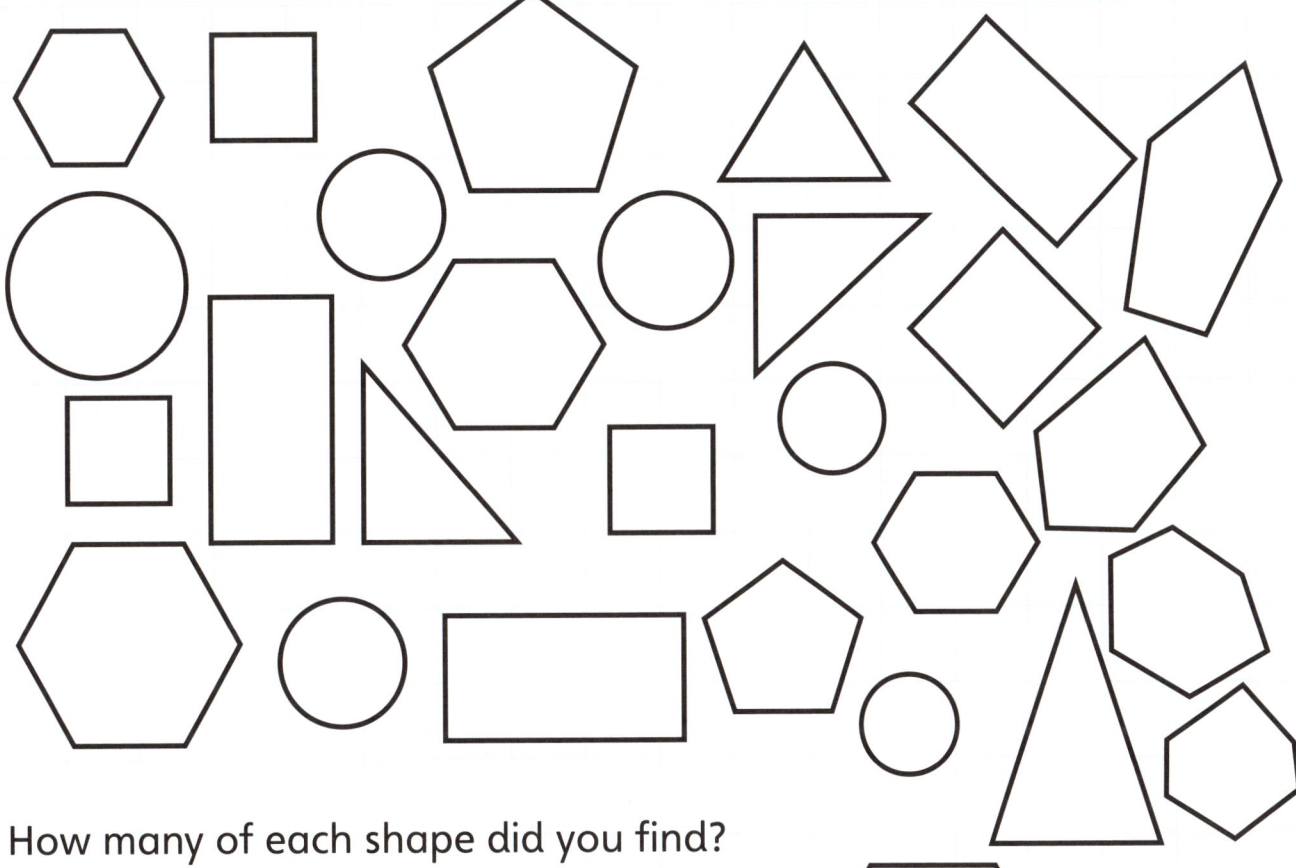

Colour the circles green, triangles blue, squares orange, rectangles purple, pentagons red and hexagons yellow. Label three of each shape with its correct name.

How many of each shape did you find?

Shape	How many?
circle	
triangle	
square	
rectangle	
pentagon	
hexagon	

Shapes Everywhere

107

Irregular shapes

Irregular pentagons have _____ sides.

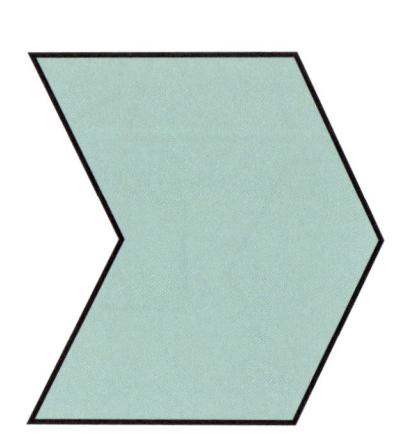

Irregular hexagons have _____ sides.

8B 3D shapes

Discover

This is a **sphere**.

It has _____ faces and _____ vertices.

This is a **cylinder**.

It has _____ faces and _____ vertices.

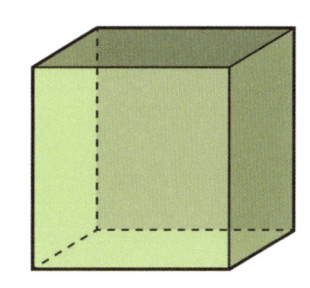

This is a **cube**.

It has _____ faces and _____ vertices.

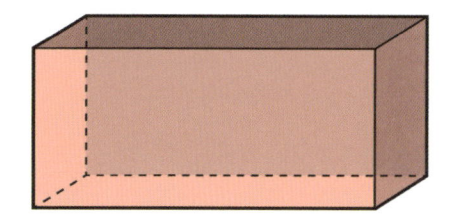

This is a **cuboid**.

It has _____ faces and _____ vertices.

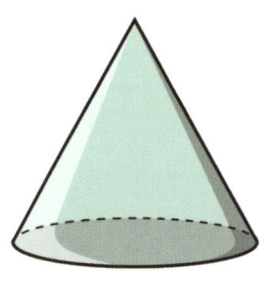

This is a **cone**.

It has _____ faces and _____ vertex.

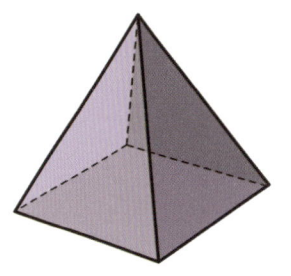

This is a **pyramid**.

It has _____ faces and _____ vertices.

8B 3D shapes

Explore

Which shapes were used to make each model?

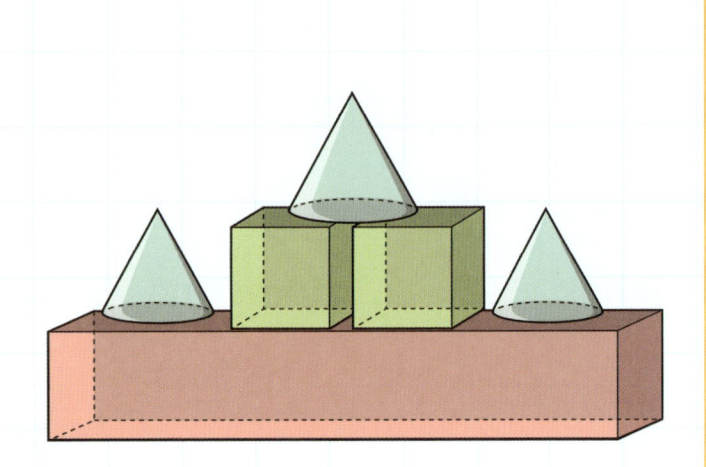

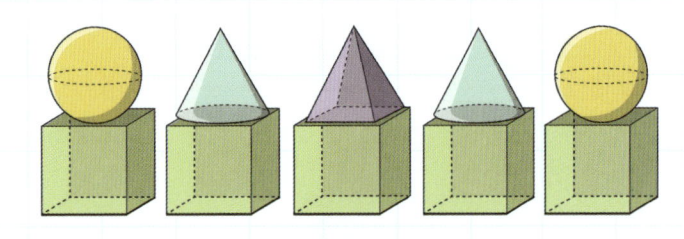

Shape	Tick ✓	How many?
Sphere		
Cylinder		
Cube		
Cuboid		
Cone		
Pyramid		

Shape	Tick ✓	How many?
Sphere		
Cylinder		
Cube		
Cuboid		
Cone		
Pyramid		

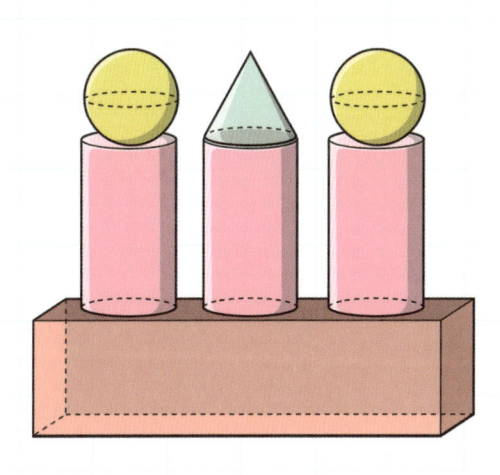

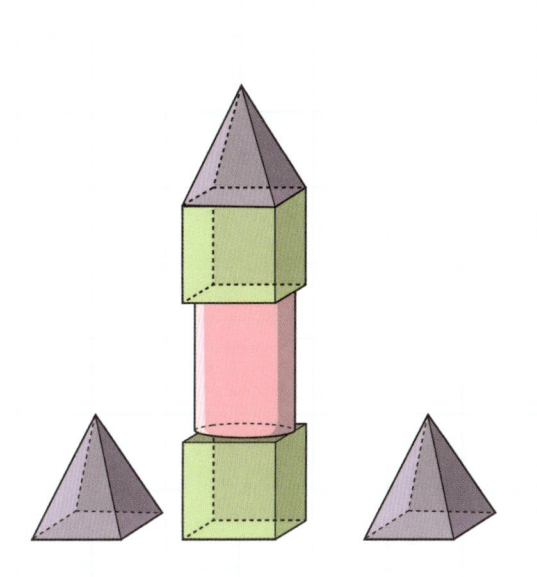

Shape	Tick ✓	How many?
Sphere		
Cylinder		
Cube		
Cuboid		
Cone		
Pyramid		

Shape	Tick ✓	How many?
Sphere		
Cylinder		
Cube		
Cuboid		
Cone		
Pyramid		

8C Sorting shapes using a Venn diagram

Discover

Write or draw the shapes in the correct places.

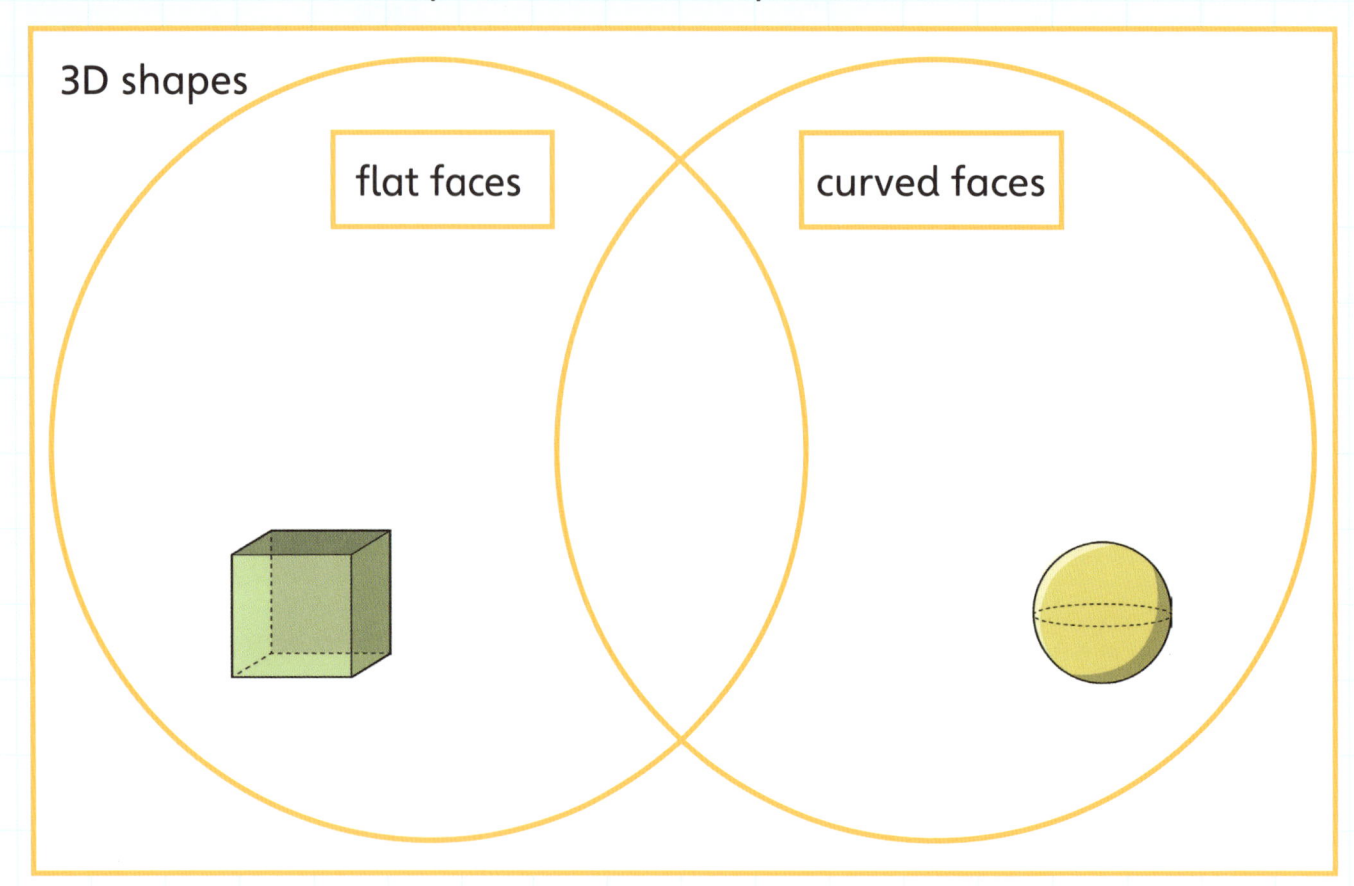

3D shapes

flat faces

curved faces

Which shapes belong in the overlap? Write their names.

_____

Use the Venn diagram to help you complete this sentence.

Shapes in the overlap have _____ and _____ edges.

112

8C Sorting shapes using a Venn diagram

Explore

Label the Venn diagram and use it to sort shapes.

First, label the whole diagram **2D shapes**, **3D shapes** or **2D and 3D shapes**.

Next, label each oval. Your labels could be a colour, a shape, the number of sides or something else.

Now sort your shapes. Either draw them or write the shape name in the correct place to record.

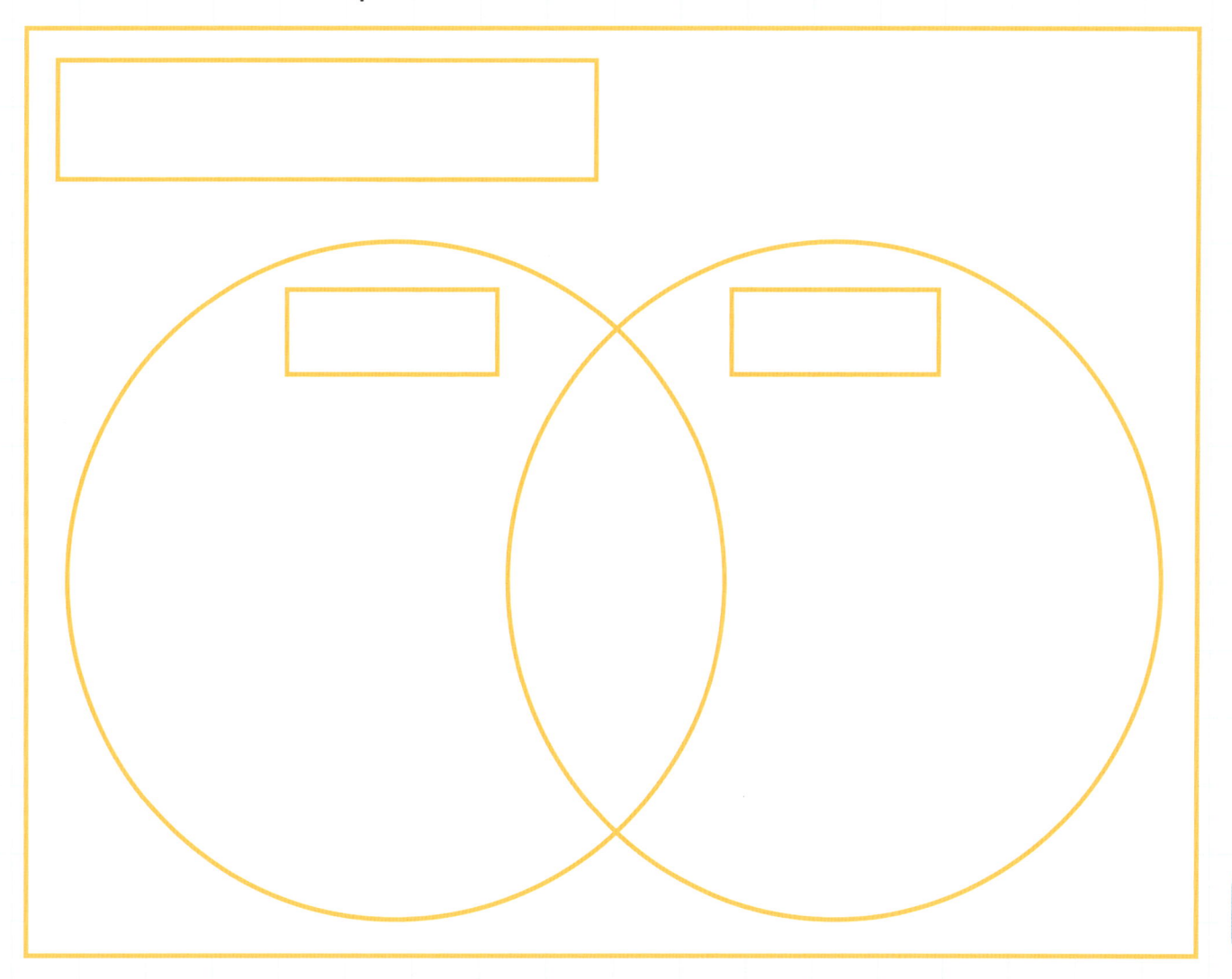

8D Sorting shapes using a Carroll diagram

Discover

	blue	not blue
More than 3 sides		
Not more than 3 sides		

114

8D Sorting shapes using a Carroll diagram

Explore

8E Reflective symmetry 1

Discover

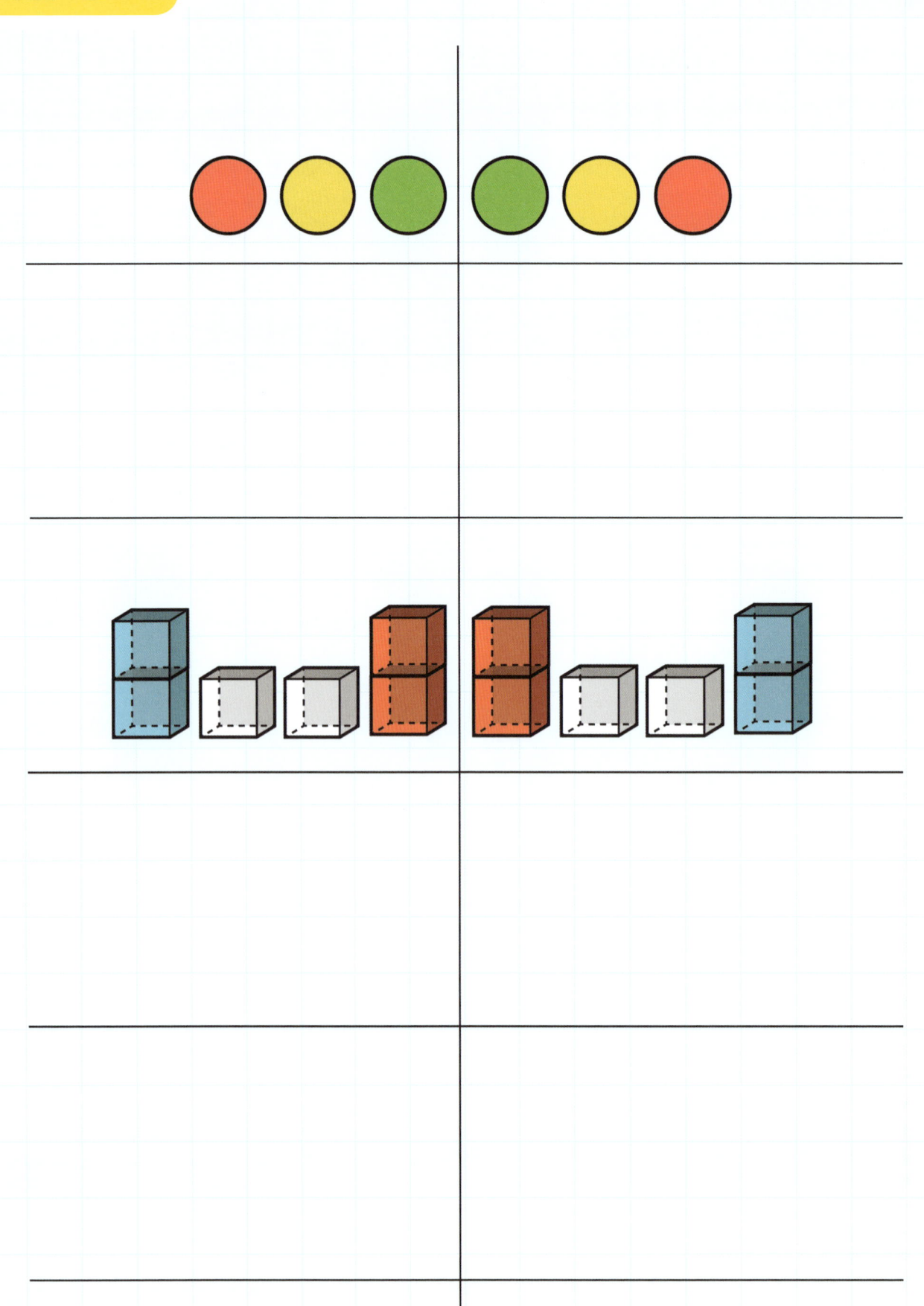

8E Reflective symmetry 1

Explore

Complete these symmetrical creatures.

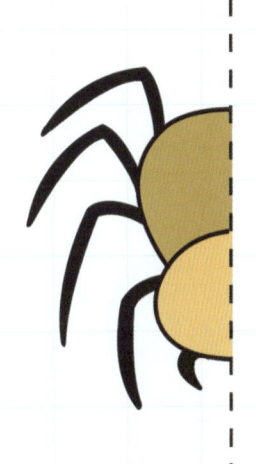

8F Reflective symmetry 2

Discover

Draw in the lines of symmetry. The first two have been done for you.

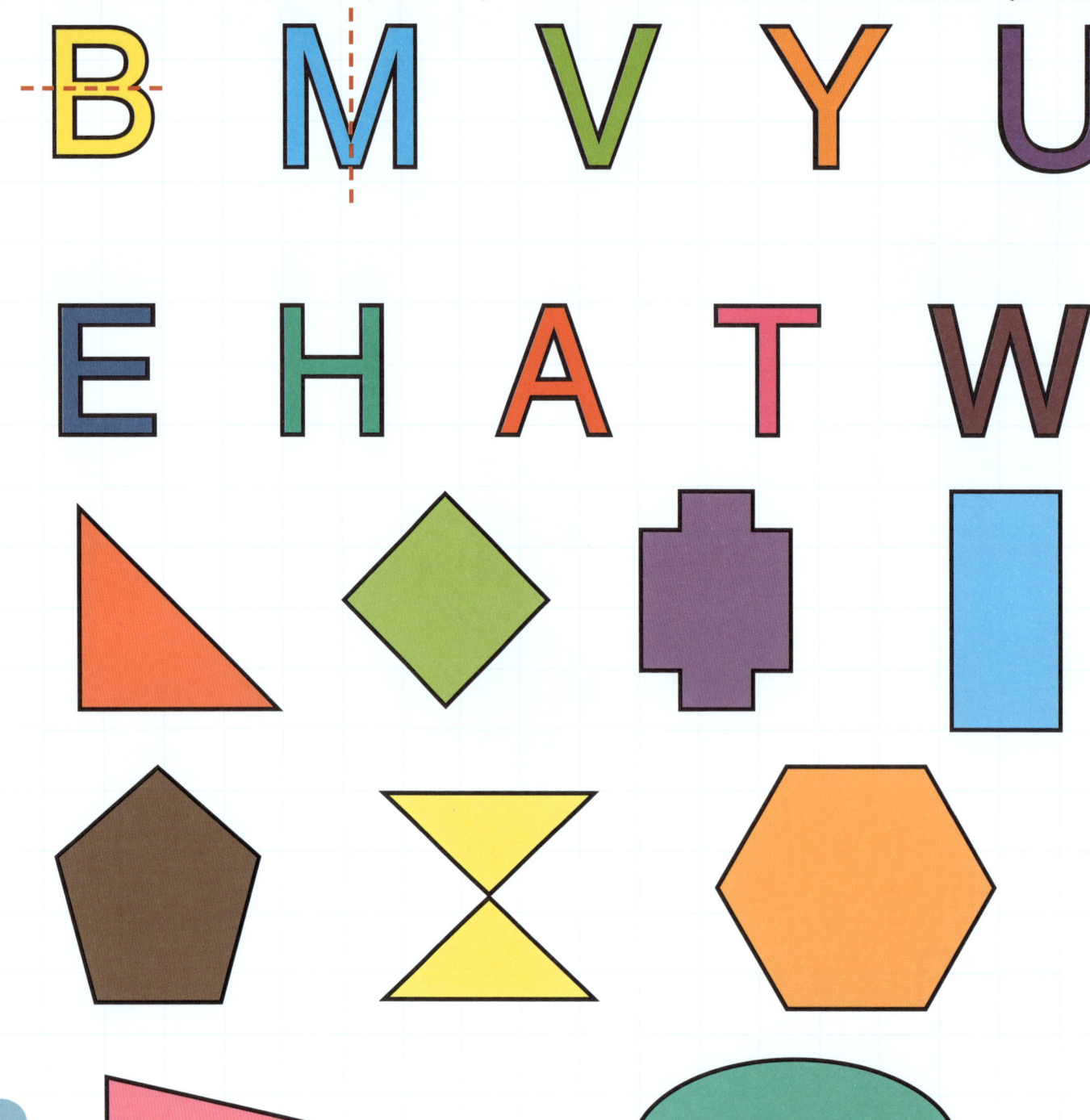

8F Reflective symmetry 2

Explore

Tick ✓ the symmetrical pictures, cross ✗ the pictures that are not symmetrical. Draw one or more lines of symmetry on the pictures you ticked.

8 Shapes everywhere

Connect

All the shapes on Planet Quirk have an odd number of sides or an odd number of faces. Which shapes will you find on Planet Quirk?

All the shapes on Planet Orb have at least one curved edge or face. Which shapes will you find on Planet Orb?

Invent a planet. What is special about the shapes on your planet?

8 Shapes everywhere

Review

True or false?

Tick ✓ the correct sentences and put a cross ✗ next to the incorrect ones.

All the sides of a square are the same length. ☐

A pyramid has 5 vertices. ☐

A pentagon has 6 straight sides and 6 vertices. ☐

Cubes have six square faces. ☐

All shapes with three straight sides are called triangles. ☐

A sphere has no vertices. ☐

All shapes with five straight sides are called hexagons. ☐

Cylinders have 4 circular faces. ☐

Cuboids have six rectangular faces. ☐

All shapes with six straight sides are called pentagons. ☐

Cones have one vertex. ☐

Choose two incorrect sentences. Re-write them to make them correct.

_____

_____

9 Measurements

Engage

How tall are you?

How are you going to do that?

You measure horses in hands, so let's do that!

I don't know, but I know I'm taller than you.

You're taller than all of us! We should measure you.

How tall are you?

Measurements

123

9A Measuring in centimetres

Discover

Object		Estimate	Measure
pencil		cm	cm
glue stick		cm	cm
crayon		cm	cm
paper clip		cm	cm
rubber		cm	cm
spoon		cm	cm
scissors		cm	cm
peg		cm	cm

Choose five more things in the classroom to estimate then measure.

Object	Estimate	Measure
	cm	cm

Complete these sentences.

The longest thing I measured was a _____ at _____ cm.

The shortest thing I measured was a _____ at _____ cm.

9A Measuring in centimetres

Explore

A _____
B _____

Which line is longer? _____

How much longer is the line? _____

A _____
B _____

Which line is shorter? _____

How much shorter is the line? _____

A _____
B _____

Which line is shorter? _____

How much shorter is the line? _____

A _____
B _____

Which line is longer? _____

How much longer is the line? _____

How long is the longest line? _____

How long is the shortest line? _____

Measure each line

A ——————————————————————————— ☐ cm

B ———————— ☐ cm

C ——————————— ☐ cm

D —————————————————————— ☐ cm

E —————————————————— ☐ cm

Use coloured pencils to draw the following lines.

Blue 15 cm   Green 4 cm   Red 7 cm   Yellow 9 cm

Purple 2 cm   Pink 13 cm   Brown 6 cm Black 12 cm

Label each line with its length.

9B String measures

Discover

Object		Estimate	Measure
wrist		cm	cm
knee		cm	cm
ball		cm	cm
table leg		cm	cm
flower pot		cm	cm
my hand		cm	cm
can		cm	cm
container		cm	cm

Choose five more things in the classroom to estimate then measure.

Object	Estimate	Measure
	cm	cm

Complete these sentences.

The longest thing I measured was a _____ at _____ cm.

The shortest thing I measured was a _____ at _____ cm.

9B String measures

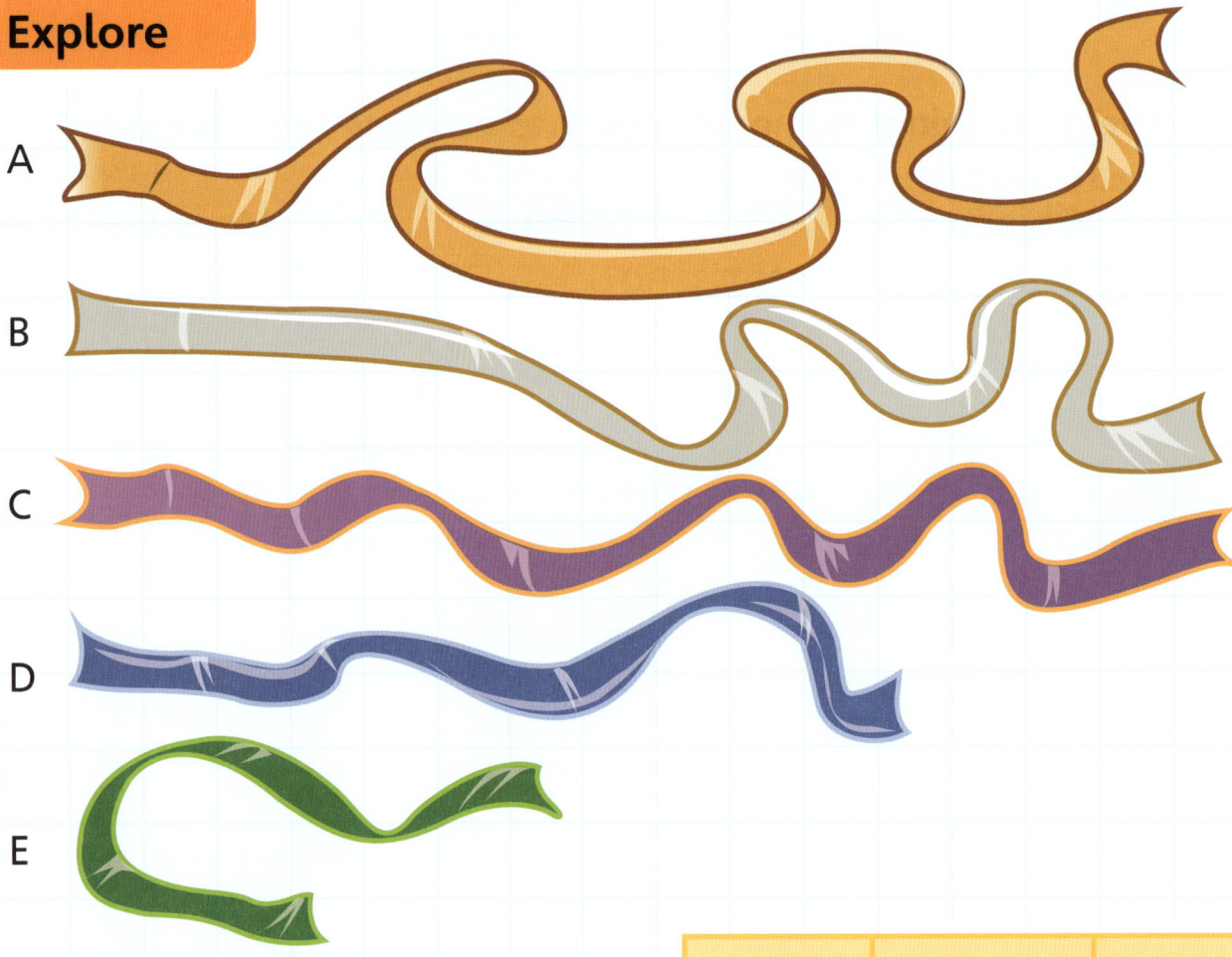

Explore

A

B

C

D

E

Use the table to help you complete these sentences.

Ribbon _____ is the longest.

Ribbon _____ is the shortest.

Ribbon _____ is 5 cm longer than Ribbon _____.

Ribbon _____ is 3 cm shorter than Ribbon _____.

Ribbon _____ is 10 cm longer than Ribbon _____.

Ribbon _____ is 14 cm shorter than Ribbon _____.

Ribbon	Estimate	Measure
A		
B		
C		
D		
E		

130

9C Metre measures

Discover

Draw or write the things you measured with your metre stick or metre mouse.

Shorter than a metre	1 metre long	Longer than a metre

How many centimetres in half a metre? _____

How many centimetres in $\frac{1}{4}$ metre? _____

How many centimetres in 2 metres? _____

How many centimetres in $1\frac{1}{2}$ metres? _____

What length is halfway between 40 cm and 50 cm? _____

What length is halfway between 65 cm and 75 cm? _____

9C Metre measures

Explore

Would you measure each of these things using centimetres or metres?

Write centimetres and metres in each box.

Length of the playground 	Width of a computer screen 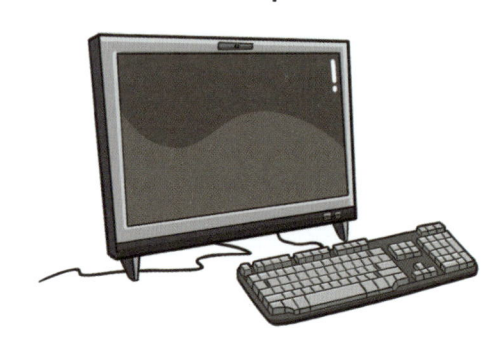
_____	_____

Length of a worm 	Length of a book cover 
_____	_____

Length of a bus 	Height of a tree 
_____	_____

Height of a dog	Length of a pencil
_____	_____
Length of a pair of scissors	Length of the classroom
_____	_____
Length of a thumb	Length of swimming pool
_____	_____

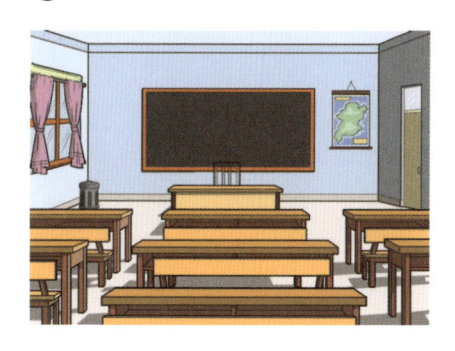

Measurements

133

9D Liquid measures

Discover

This jug holds 1 litre of water.

Your teacher will show you some containers. Use a litre jug to help you decide if each container will hold less than a litre or more than a litre.

Holds less than a litre	Holds more than a litre

bucket	 watering can	 fish tank	 mug
 teaspoon	can of fizzy drink	large bottle of fizzy drink	 vase
 teapot	carton of drink	 cereal bowl	1 litre jug

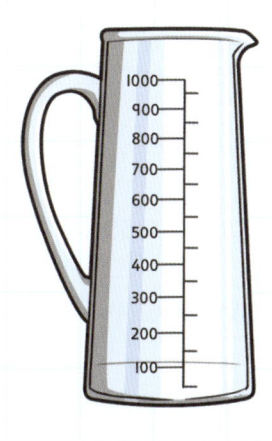

Measurements

135

9D Liquid measures

Explore

Lemonade recipe

1 litre water

4 lemons

100 g sugar

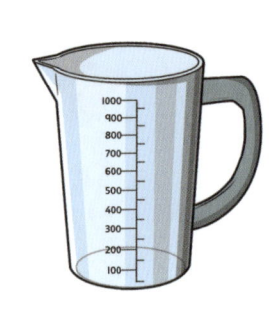

1 litre = 1000 ml

How many millilitres in half a litre? _____

How many millilitres in a quarter of a litre? _____

1 litre of lemonade is poured into 5 glasses.

Each glass has the same amount.

How many millilitres in each glass? _____

1 litre of lemonade is poured into 4 glasses.

Each glass has the same amount.

How many millilitres in each glass?

_____

Show how you work out the answers to these questions.

I have 20 lemons. How many litres of lemonade can I make?

_____

How many 500 ml bottles can I fill? _____

How many 200 ml drinks will I have? _____

9E Sweet cake weights

Discover

Grandma's simple sweet cake recipe
1 egg
100g self-raising flour
100g butter
100g sugar
Makes 6 small cakes

1 large egg weighs 100g. How many of each object weighs the same as an egg?

		Heavier than an egg
cubes		
counters		
paperclips		
pencils		

138

Write or draw the objects that weigh the same as an egg in order from lightest to heaviest.

lightest
heaviest

9E Sweet cake weights

Explore

Grandma's simple sweet cake recipe
1 egg
100g self-raising flour
100g butter
100g sugar
Makes 6 small cakes

A large egg weighs 100 g.

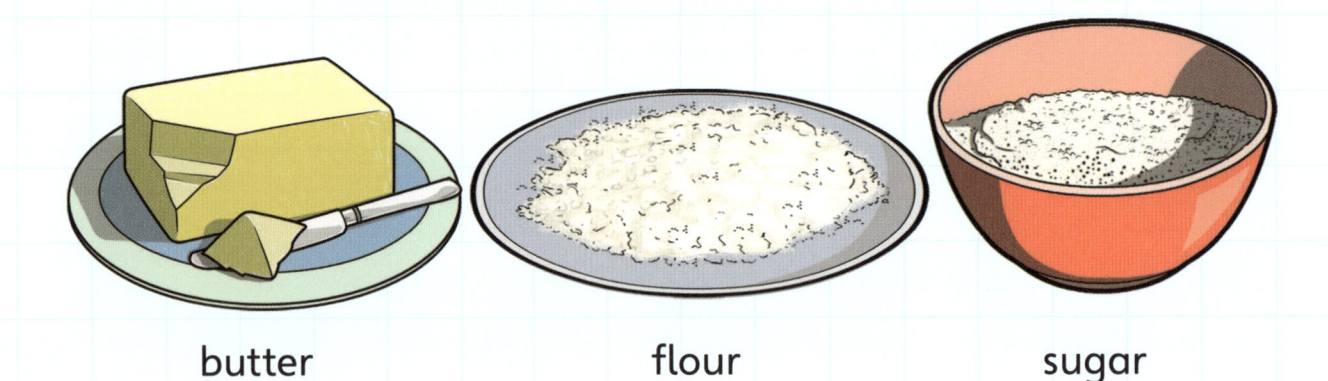

butter   flour   sugar

Complete the missing amounts.

1 egg

_____ g butter

_____ g self-raising flour

_____ g sugar

Makes 6 small cakes

_____ eggs

200 g butter

200 g self-raising flour

200 g sugar

Makes _____ small cakes

5 eggs

_____ g butter

_____ g self-raising flour

_____ g sugar

Makes _____ small cakes

_____ eggs

1 kg butter

1 kg self-raising flour

1 kg sugar

Makes _____ small cakes

Look at the labels on some food packaging to find the weight of the contents. Draw or write the item and its weight.

9F Money

Discover

Which coins could you use to pay for each toy?

43 c	
18 c	
75 c	
82 c	

55 c	
77 c	
28 c	

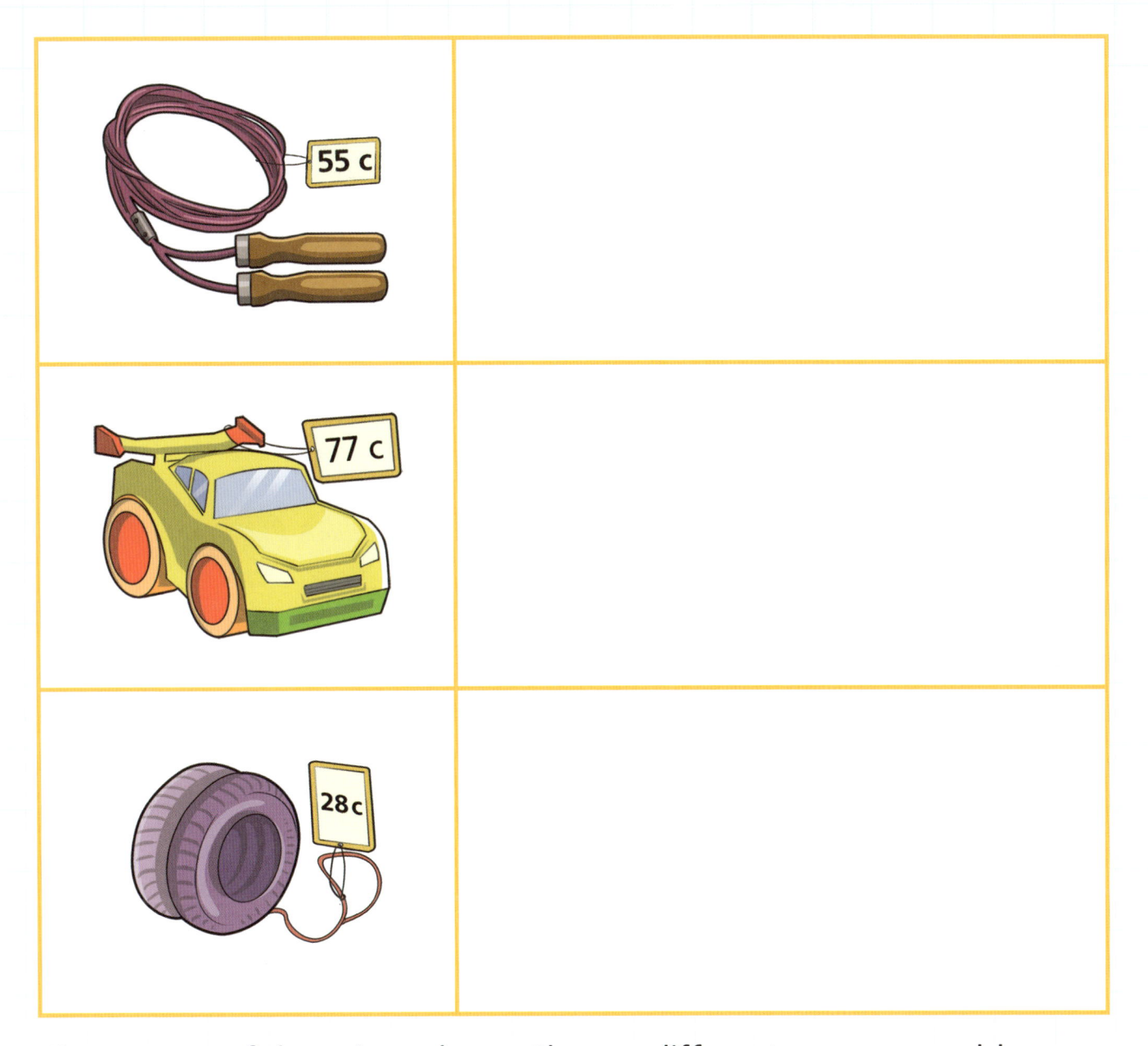

Choose one of the prices above. Show a different way you could pay.

9F Money

Explore

You buy each toy and pay with $1.
Use the number line to help you work out your change.
Draw the coins you might get. Remember, 100c = $1

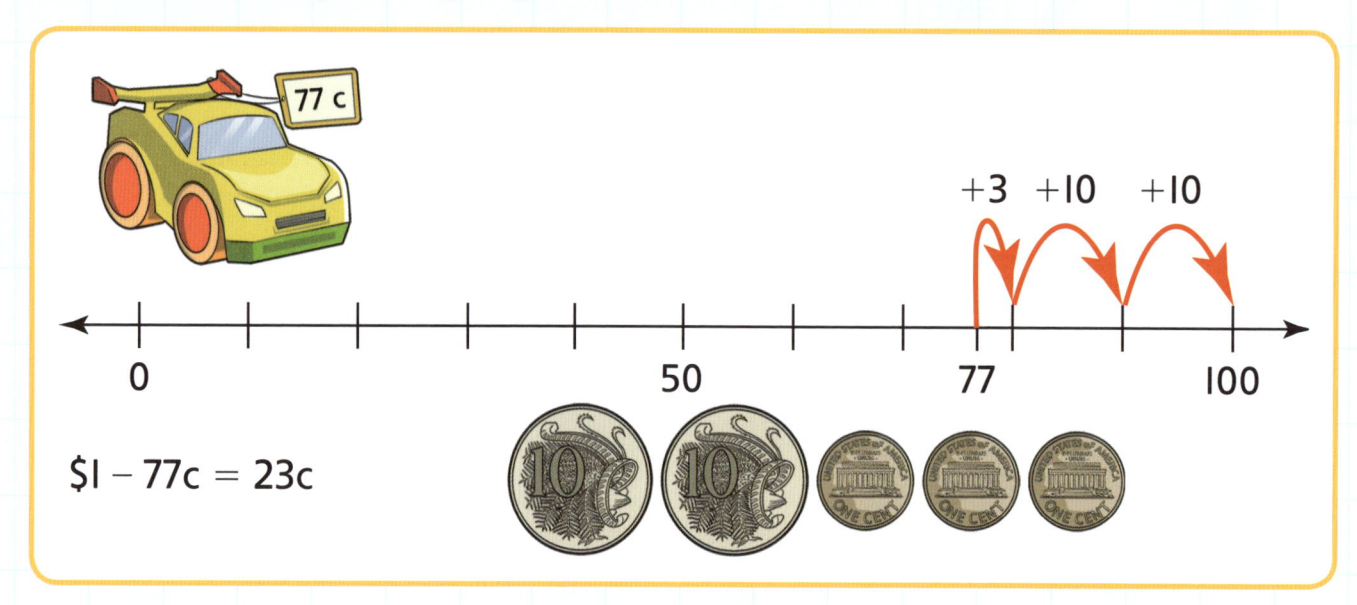

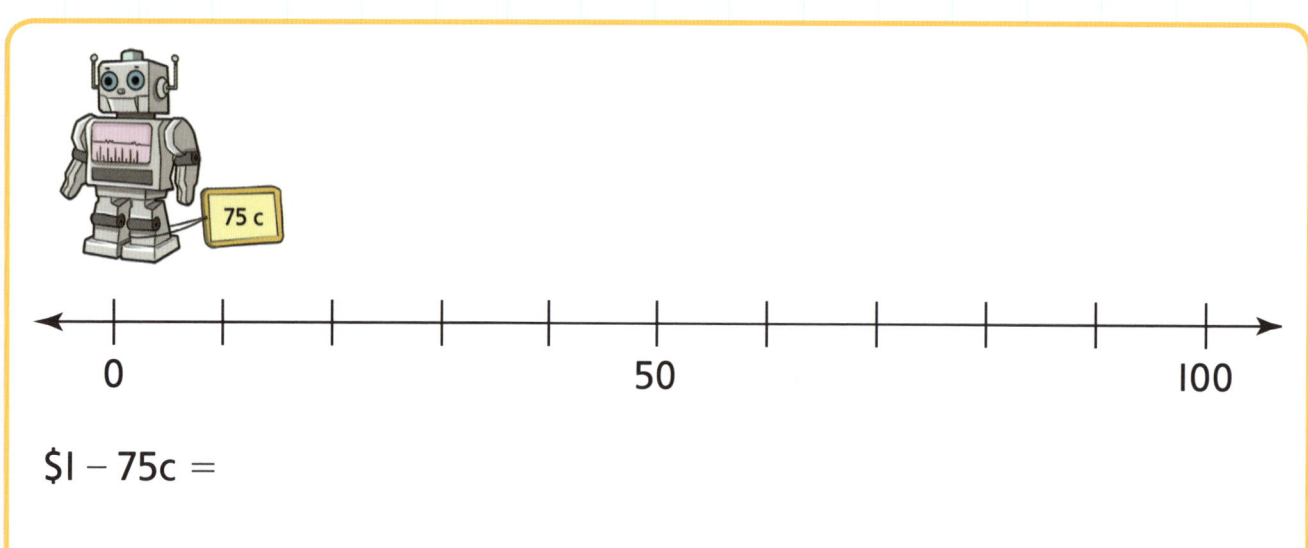

77 c

+3 +10 +10

0   50   77   100

$1 − 77c = 23c

75 c

0   50   100

$1 − 75c =

82 c

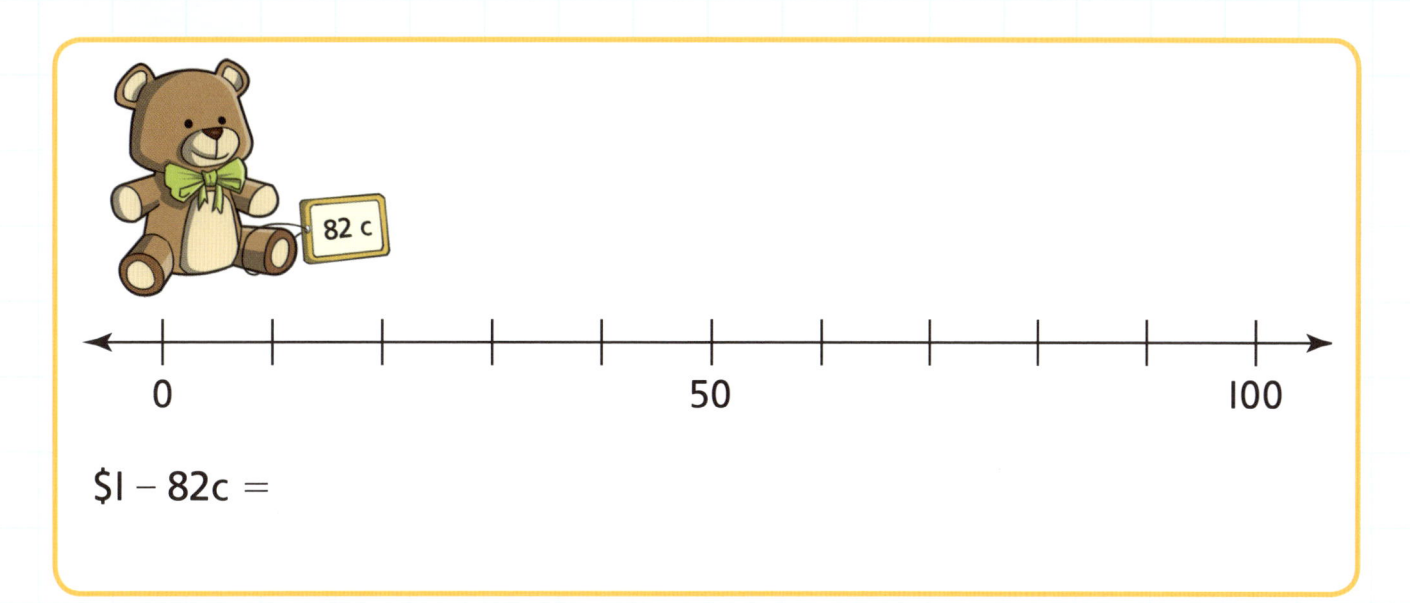

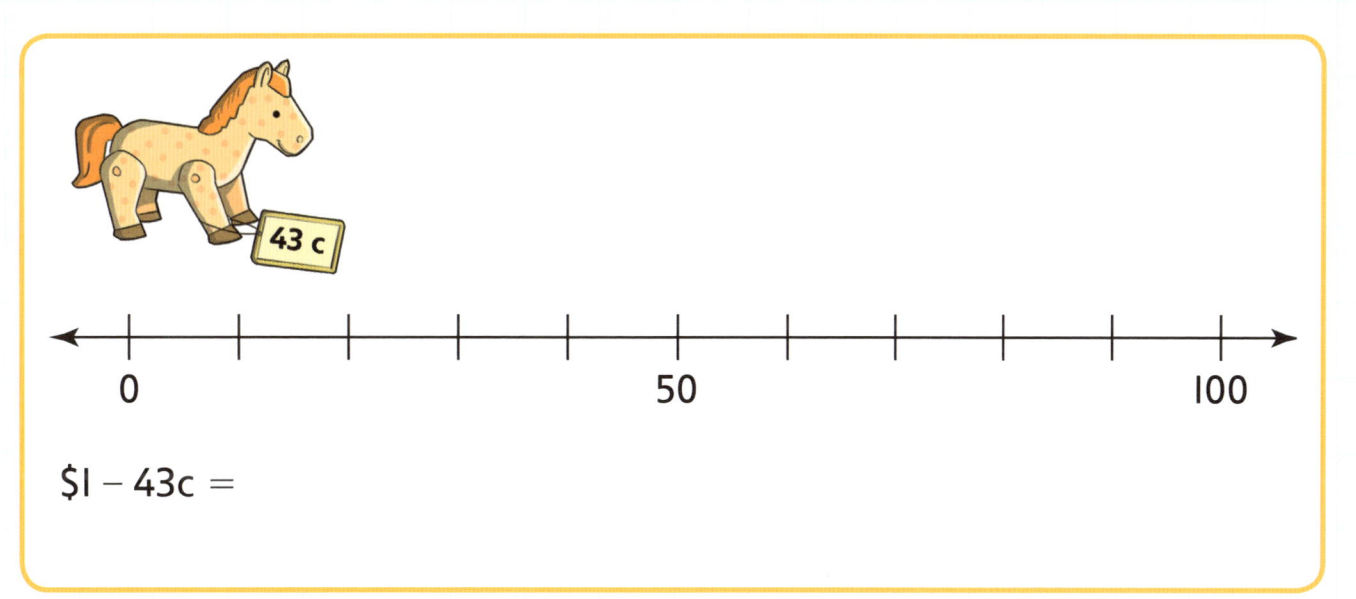

0   50   100

$1 − 82c =

43 c

0   50   100

$1 − 43c =

Make up a story to go with one of the calculations. Here is an example.

My mother gave me $1. We went to the shops and I bought a toy car for 77c. I got 23c change.

_____

_____

9 Measurements

Connect

For your class party, each student will get two glasses of lemonade, a sweet cake and a flapjack.
Everyone will also get a 50 g bag of fudge to take home.

Sweet cake recipe

(makes 6 small cakes)

1 egg

100 g self-raising flour

100 g butter

100 g sugar

Till receipt

1 kg sugar	$2.00
500 g butter	$2.00
6 eggs	$1.20
1 kg flour	$1.00

Flapjacks

(makes 20)

200 g rolled oats

50 g caster sugar

100 g butter

100 g golden syrup

Till receipt

1 kg oats	$2.00
1 kg sugar	$2.00
500 g butter	$2.00
500 g syrup	$1.50

Lemonade recipe

1 litre water

4 lemons

100 g sugar

1 litre = 1000 ml

Till receipt

Pack of 4 lemons	$1.00
1 kg sugar	$2.00

Simple fudge

(makes 500 g)

50 g butter

400 g sugar

125 ml of milk

2½ ml spoon vanilla essence

Till receipt

1 kg sugar	$2.00
500 g butter	$2.00
2 l milk	$0.48
100 ml vanilla essence	$4.00

How much will these cost to make?

Write out your shopping list and work out how much you need to spend to get everything you need.

	1 kg	500 g	100 g	50 g
Oats	$2	$1	20c	
Sugar				10c
Butter	$4		40c	
Syrup		$1.50		15c

	2 litres	1 litre	500 ml	250 ml	125 g
Milk	48c				

	100 ml	1 ml	2½ ml
Vanilla essence	$4.00	4c	

9 Measurements

Review

Metre Mouse has a tail 1 metre long. 1 m = 100 cm

How long is a Half Metre Mouse tail? ☐

How long is a Quarter Metre Mouse tail? ☐

How long is a Three-Quarter Metre Mouse tail? ☐

How long is a Two-Quarter Metre Mouse tail? ☐

Two mice put their tails end to end. If they are a Half Metre Mouse and a Two-Quarter Metre Mouse, how long are the tails altogether? ☐

A large egg weighs 100 g. I follow Grandma's Simple Sweet Cake recipe using 2 eggs.

How much does the cake mixture weigh altogether? ☐

Today I drank three mugs of tea, a 500 ml bottle of water and two glasses of lemonade. My mug holds 300 ml and my glass holds 200 ml. How much liquid did I drink today?

10 Geometry

Engage

Cross the road by the park.

How do you get to school?

Follow the path between the two blocks of flats.

left at the crossroads

turn right at the shop

149

10A Turns and right angles

Discover

Make a right angle checker:

Fold a circle of card in half.

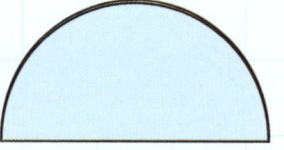

Fold it in half again to make a right angle.

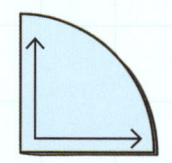

Use your right angle checker to test for right angles.

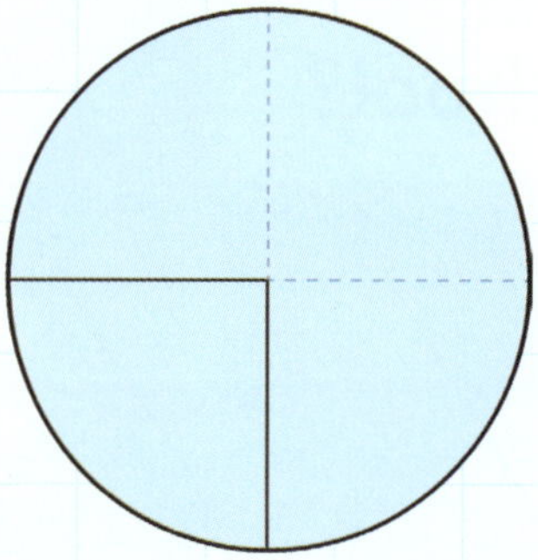

Tick ✓ the right angles.

10A Turns and right angles

Explore

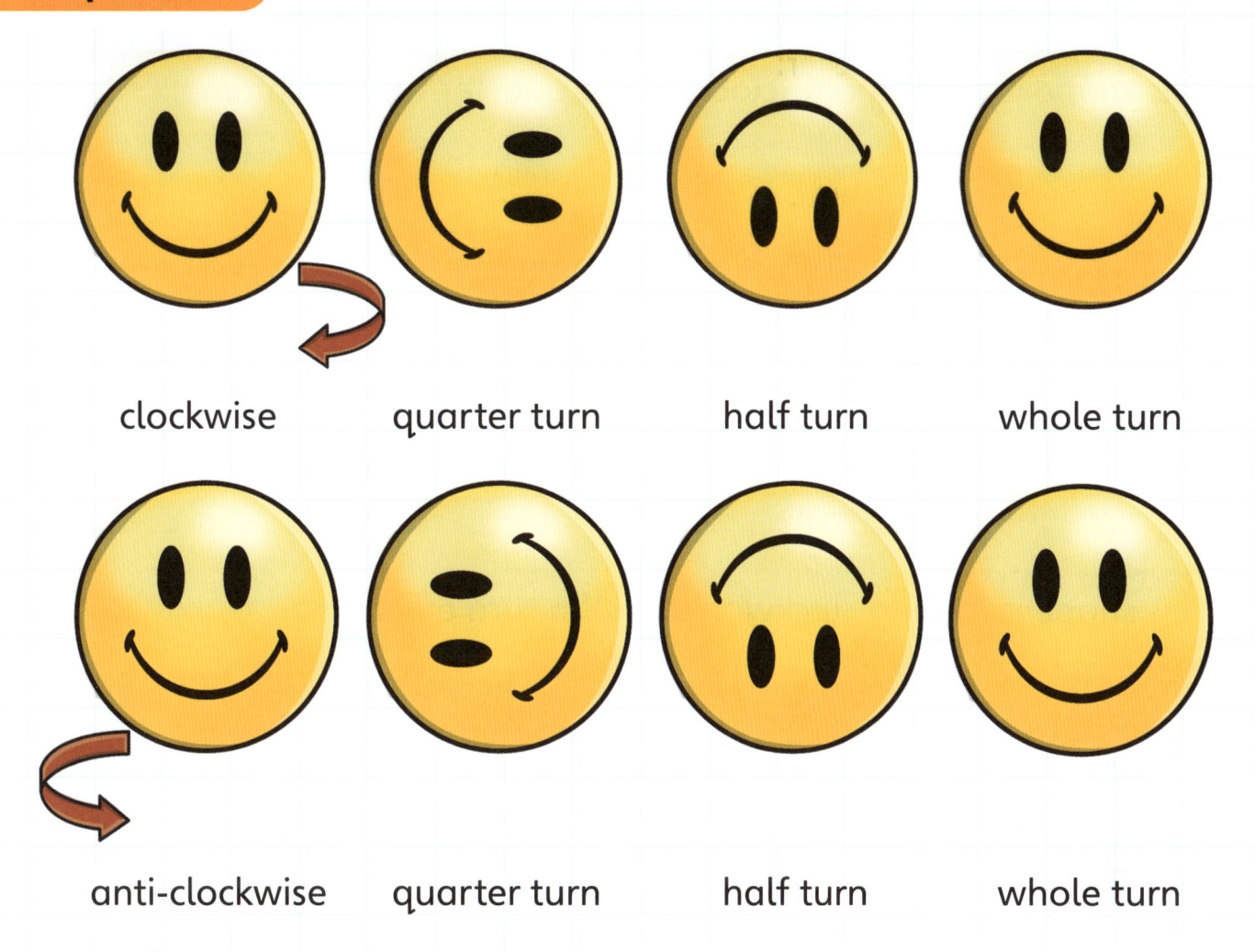

clockwise | quarter turn | half turn | whole turn

anti-clockwise | quarter turn | half turn | whole turn

Draw the next two shapes in each pattern.

This shape is turning clockwise, a quarter turn each time.

Geometry

151

This shape is turning anti-clockwise, a quarter turn each time.

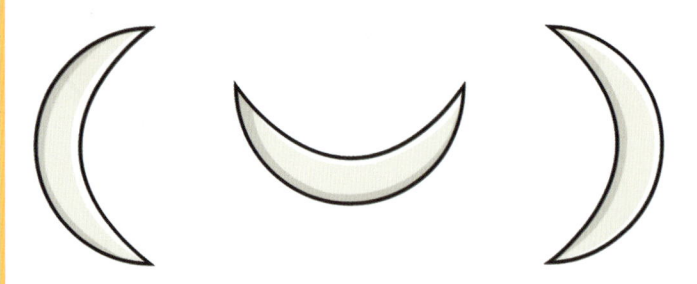

This shape is turning clockwise, a half a turn each time.

This shape is turning anti-clockwise, a quarter turn each time.

152

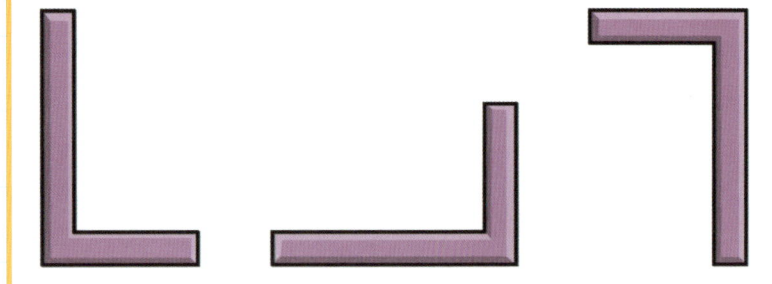

10B Travelling

Discover

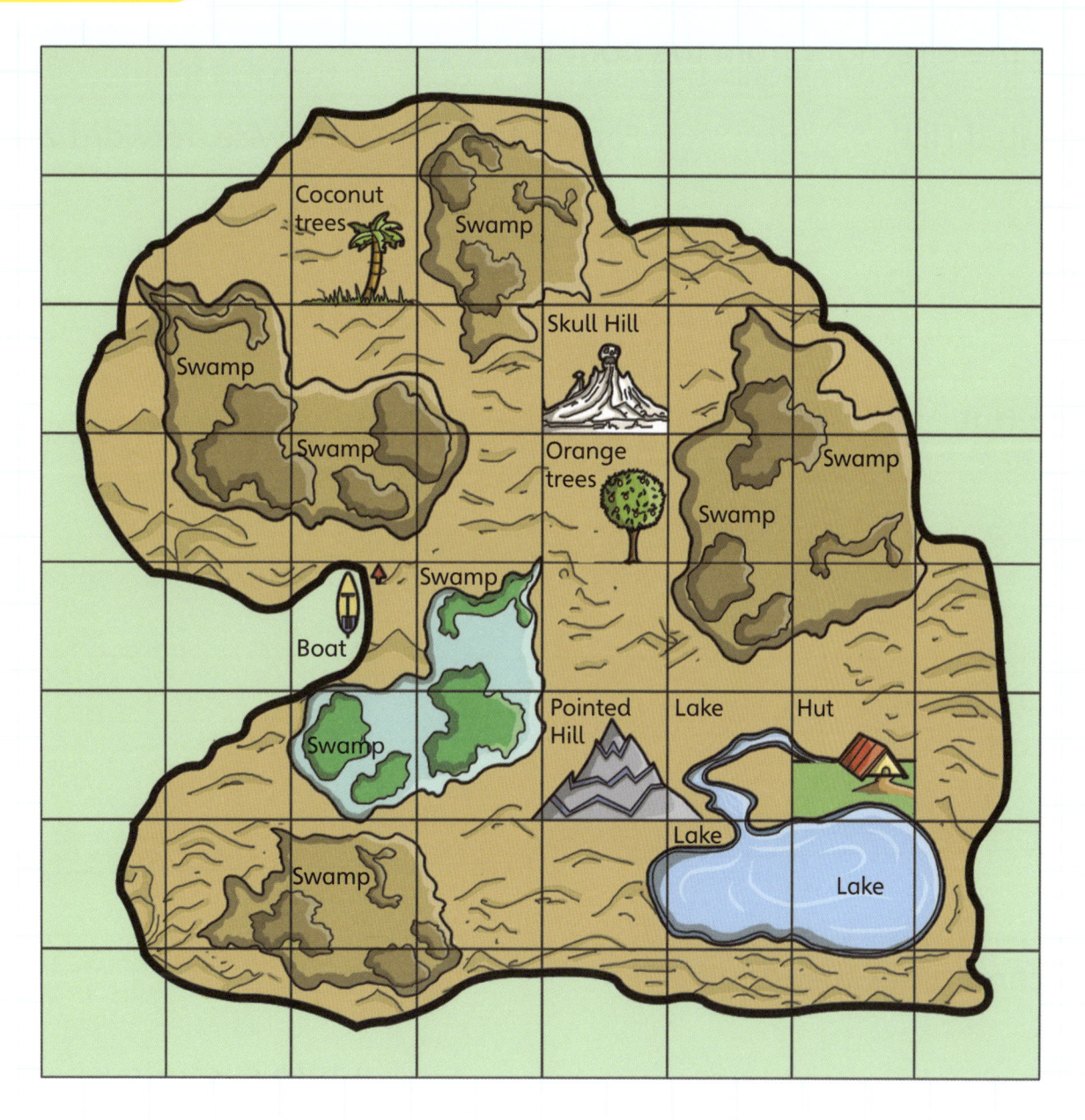

Travel advice

- Travel from the boat in the direction shown.
- You cannot go through the swamp or lake. You must go around them.
- Always face the direction you are travelling.
- No diagonal moves.

Write the directions from the boat to:

Pointed Hill	Forward 1, $\frac{1}{4}$ turn clockwise, forward 2, $\frac{1}{4}$ turn clockwise, Forward 2. F1, $\frac{1}{4}$ C, F2, $\frac{1}{4}$ C, F2.
The Hut	
Orange trees then Coconut trees	
Coconut trees then Skull Hill	
Skull Hill then the Hut	

10B Travelling

Explore

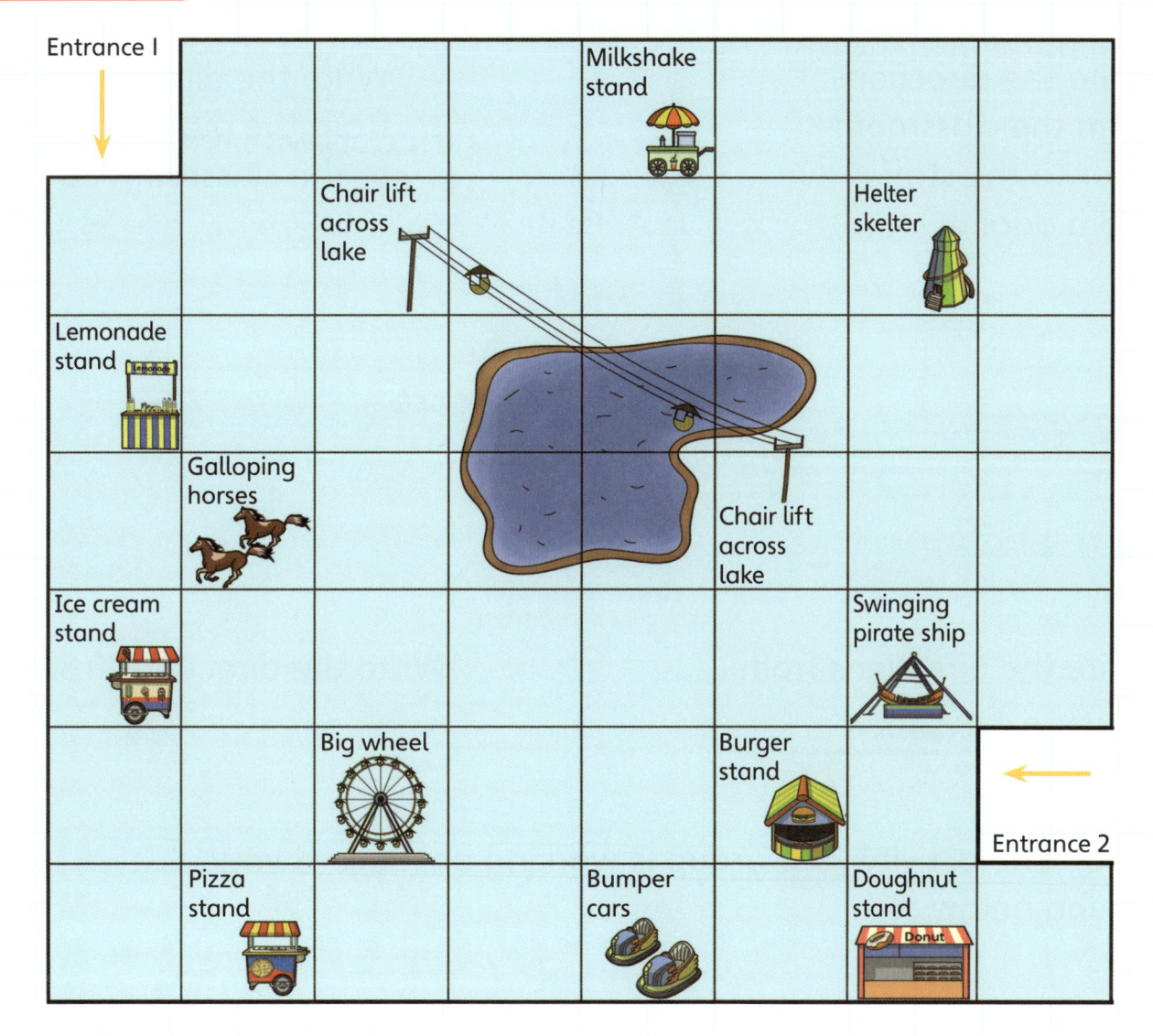

Travel advice

- Always face the direction you are travelling.
- No diagonal moves.

Which entrance will you use to get into the fun fair?
Will you leave from the same entrance?

Write your journey around the fun fair. Make sure you visit most of the rides and stalls. Don't forget to get something to eat and drink.

Geometry

155

10 Geometry

Connect

Write the directions from the classroom door to the staff room door.

Write the directions from the playground to the classroom door.

Write the directions from the front entrance of the school to the classroom door.

Write the directions from

Write the directions from

Tick ✓ the activity your group is working on. Record your thinking below.

10 Geometry

Review

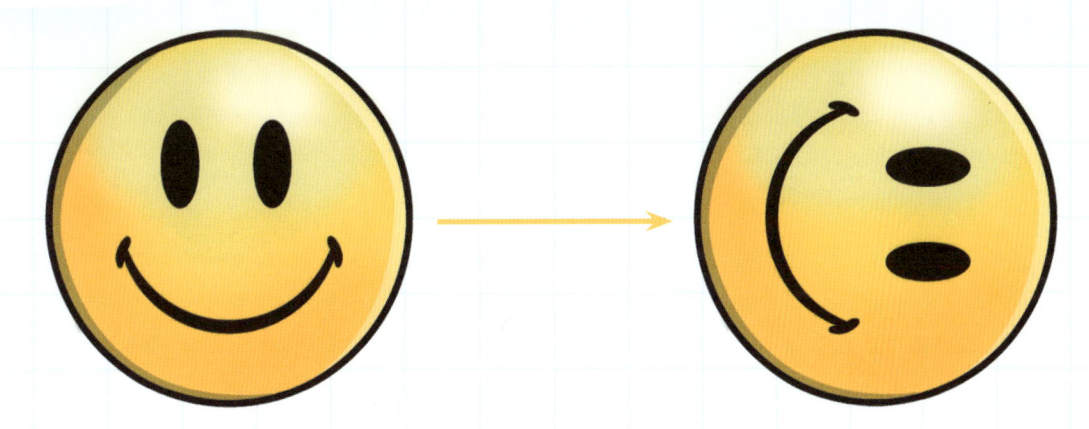

Write two different ways that the smiley face could be turned to its new position.

You might find some of these words helpful:

clockwise, anti-clockwise, quarter turn, half turn, whole turn

_____

_____

_____

Tick ✓ the right angles in these shapes.

11 Time

Engage

11A Seconds and minutes

Discover

1 minute challenges

How many cubes can you make into a tower? ☐

How many coins can you stack into a tower? ☐

Start from 1 and write the numbers in order.
Which number can you write up to in a minute?

Estimate how long it will take you to do each of these activities, then ask your partner to time you.

Activity	Estimated time in seconds	Actual time in seconds
Count to 100		
Say the alphabet		
Say the 2 times table, to 10 × 2		
Throw and catch a ball		

Complete these sentences.

There are _____ seconds in a minute.

There are _____ seconds in half a minute.

There are _____ seconds in a quarter of a minute.

11A Seconds and minutes

Explore

Sort these activities into ones that take seconds and ones that take minutes.

- brush your teeth
- yawn
- take a shower
- clap
- make toast
- play football
- hop
- eat a meal
- write your name
- blow your nose
- blink
- sneeze
- drink a glass of milk

seconds	minutes

11B Half past

Discover

Join the clocks which show the same time.

Circle the clock that does not have a partner.

11B Half past

Explore

Complete the missing times.

If it is half past 2 now, what time will it be in half an hour's time? Complete the digital clock to show the time.

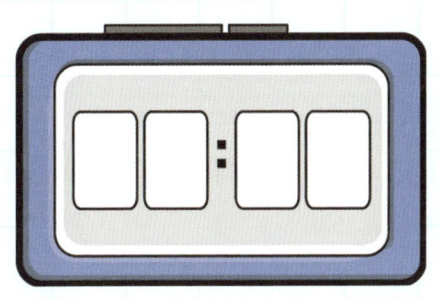

11C Days of the week

Discover

Put the days of the week in order, starting with Monday.

M
T
W
T
F
S
S

Word Bank
Wednesday   Monday
Sunday   Saturday
Friday   Thursday
Tuesday

What day is it today?

_____

What day was it yesterday?

_____

Two different days begin with S. Which day comes first in the week?

_____

Two different days begin with T. Which day comes later in the week?

_____

166

Which day comes after Thursday?

_____

Which day comes before Wednesday?

_____

What is the 3rd day of the week?

_____

What is the 5th day of the week?

_____

> Which is your favourite day of the week?
> Write or draw the reason why.

11C Days of the week

Explore

Days of the week game

Game 1

Player 1   Player 2

Game 2

Player 1   Player 2

Complete these sentences.

There are _____ days in a week.

Today is _____.

Tomorrow will be _____.

11D Months of the year

Discover

Put the months of the year in order.

Write how many days are in each month.

J	31
F	
M	
A	
M	
J	
J	
A	
S	30
O	
N	
D	

Word Bank

August   November
April   December
May   February
March   January
June   September
July   October

What is the 4th month of the year?

_____

What is the 9th month of the year?

_____

What is the 3rd month of the year?

_____

Complete the grids with the missing months.

April		June	

January			April

	October		December

June		August	

	December	January	

170

11D Months of the year

Explore

Use the word bank and your months of the year wheel to help you answer these questions.

Word Bank

August	November
April	December
May	February
March	January
June	September
July	October

Which month follows June?

_____

Which month is just before May?

_____

Which month follows March?

_____

Which month is two months after August?

_____

Which month is two months before December?

_____

Three different months begin with J. Which month comes earliest in the year?

_____

Two different months begin with M.
Which month comes later in the year?

_____

Two different months begin with A.
Which month comes earlier in the year?

_____

Which month is it now?

_____

Which month was it before this one?

_____

Which month will it be after this one?

_____

Which is your favourite month of the year?
Write or draw the reason why.

11 Time

Connect

> How many minutes are you at school for each day?

> When is (or was) the 100th day of school this year?

> When is (or was)...

> How many.....

> When is (or was) the 50th day of school this year?

> How many days have you been at school for so far this year?

Tick ✓ the activity your group worked on.

Record your thinking below. What did you find out?

11 Time

Review

Write the days of the week and the months of the year in order. Choose which day and month to start with.

Months of the year	Days of the week

Show the time on both clocks.

Half past 4

Complete these sentences.

There are _____ seconds in a minute.

There are _____ minutes in an hour.

There are _____ minutes in half an hour.

12 Handling Data

Engage

Last week this vending machine sold 24 cans of lemonade, 27 cans of cola, 23 cans of orangeade, 21 cans of fruit soda, 15 bottles of sparkling water and 13 bottles of still water.

12A Block graphs and pictograms

Discover

Complete the frequency chart and block graph for favourite breakfast cereals.

Favourite Breakfast Cereal

Cereal	Number of votes	Frequency
Cornflakes	|||| ||||	
Chocco pops	|||| |||| ||||	
Wheats	|||| ||	
Crispy rice	|||| |||| ||	
Fruity bran flakes	||||		

Favourite Breakfast Cereal

176

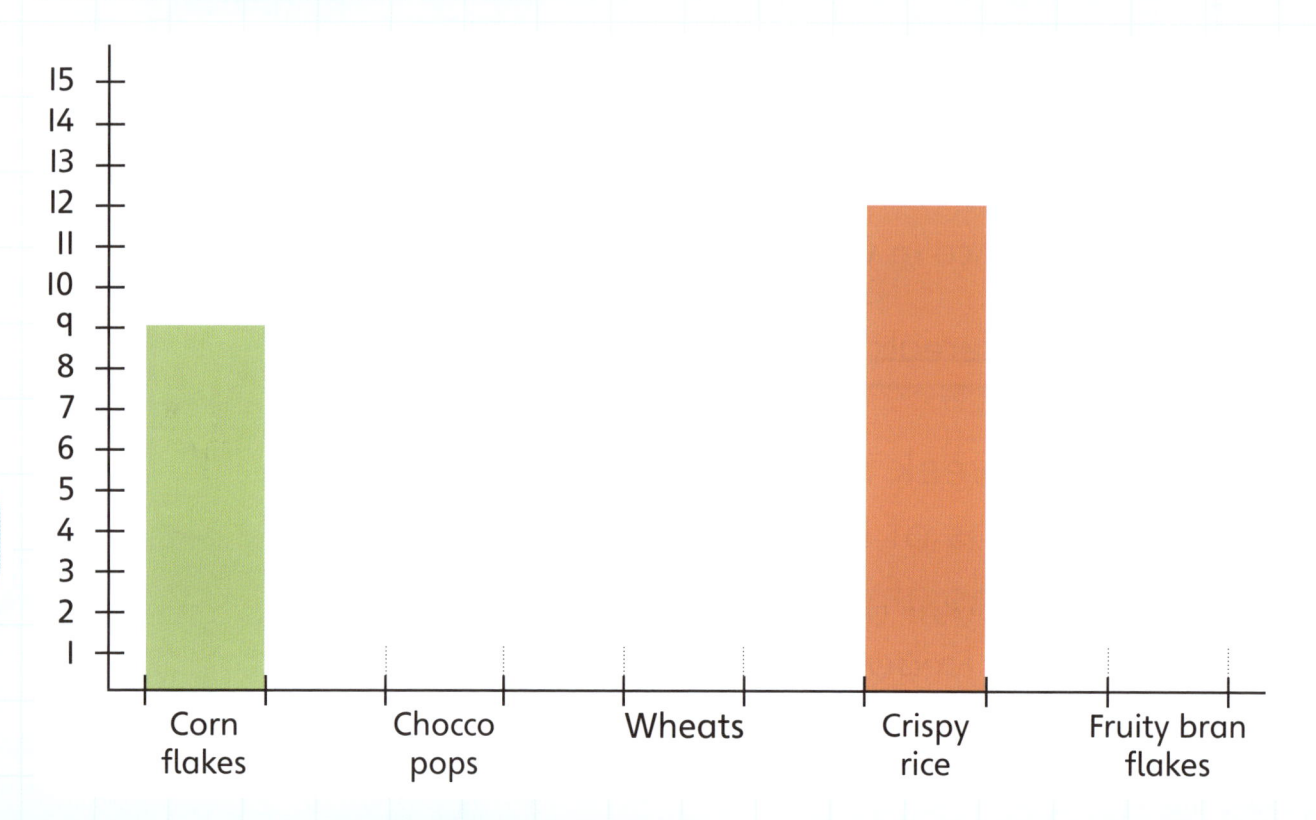

Each student had two votes. How many students were asked? ☐

What is the most popular cereal? _____

What is the least popular cereal? _____

How many students preferred Crispy rice? ☐

How many students preferred Chocco pops? ☐

Which cereal received 7 votes? _____

Which cereal received 5 more votes than that? _____

How many more students like Chocco Pops than like Fruity bran flakes? ☐

How many more students like Cornflakes than like Wheats? ☐

Which two cereals received 19 votes altogether?

_____

What is the difference in votes between the most popular and least popular cereals? ☐

How do you know?

_____

The three most popular cereals will be offered on the breakfast club menu. Which cereals are they?

_____

Handling Data

177

12A Block graphs and pictograms

Explore

Use the information in the block graph to help you complete the pictogram.

Favourite Breakfast Drink

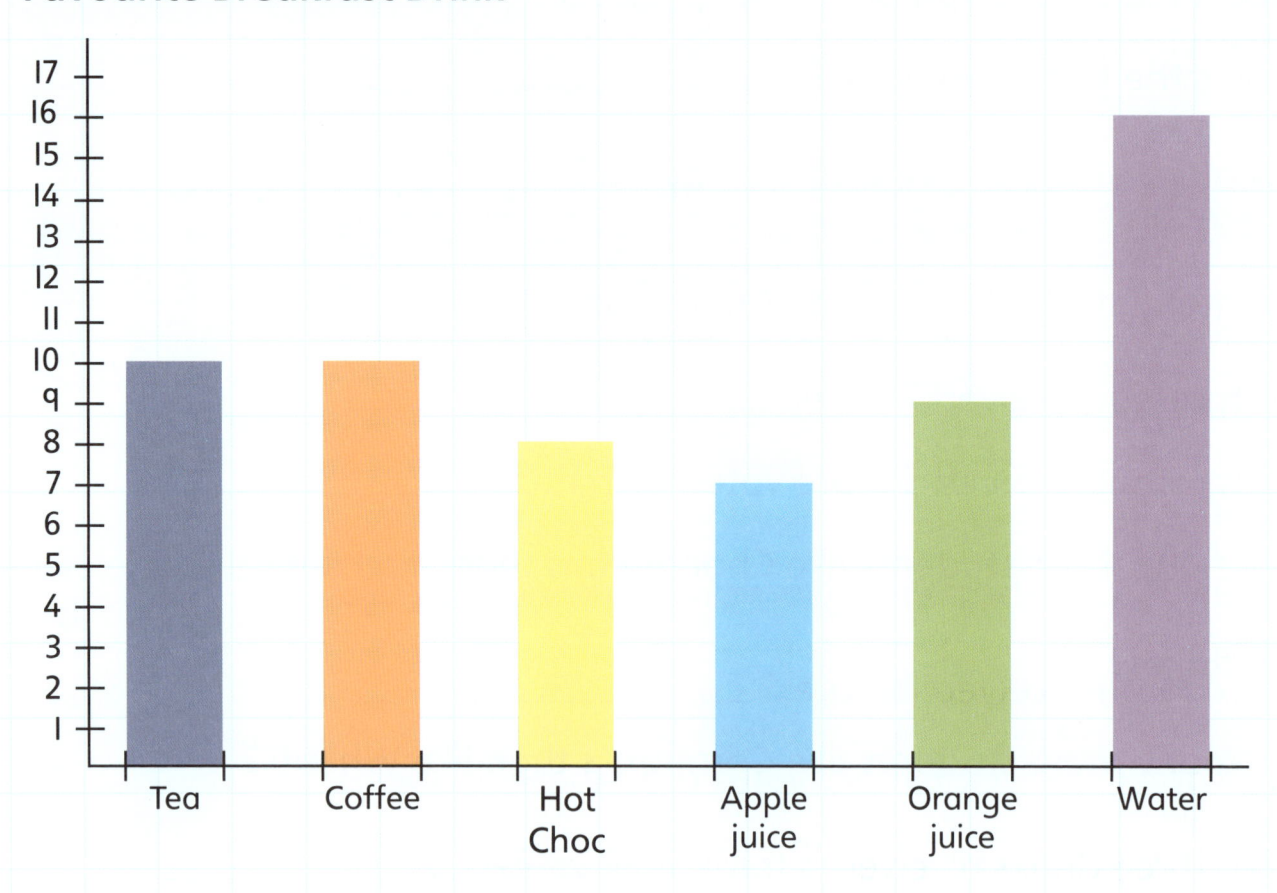

Favourite breakfast drink pictogram

Tea	😊	😊	😊	😊	😊	😊	😊	😊	😊	😊							
Coffee																	
Hot choc																	
Apple juice																	
Orange juice																	
Water																	
	1	2	3	4	5	6	7	8	9	10	11	12	13	14	15	16	17

Key 😊 = 1 vote

Each student was asked to choose their two favourite drinks at breakfast time. How many students were asked? ☐

What is the most popular drink? _____

What is the least popular drink? _____

How many students preferred hot chocolate? ☐

How many students preferred orange juice? ☐

How many more students like coffee than like apple juice? ☐

How many more students like water than like apple juice? ☐

Which drinks received 10 votes? _____

Which drink received 6 more votes than these drinks? _____

Which three drinks received a total of 24 votes? _____

Which two drinks received a total of 24 votes? _____

Which was more popular, hot drinks or cold drinks? _____

The three most popular drinks will be on the breakfast club menu.

Which drinks are they? _____

After the breakfast club opened, students complained that water was the only cold drink on the menu. Which other cold drink could be offered? Why?

_____

12B Sorting using Carroll diagrams

Discover

Complete the Carroll diagram.

Complete this sentence.

I sorted numbers for _____ and _____ in the Carroll diagram.

Complete this sentence using the words **do** or **do not**.

When you cross out the sorting title, it means that those numbers _____ match the sorting title.

12B Sorting using Carroll diagrams

Explore

Complete the Carroll diagram.

12C Sorting using Venn diagrams

Discover

Write or draw the information in the correct places.

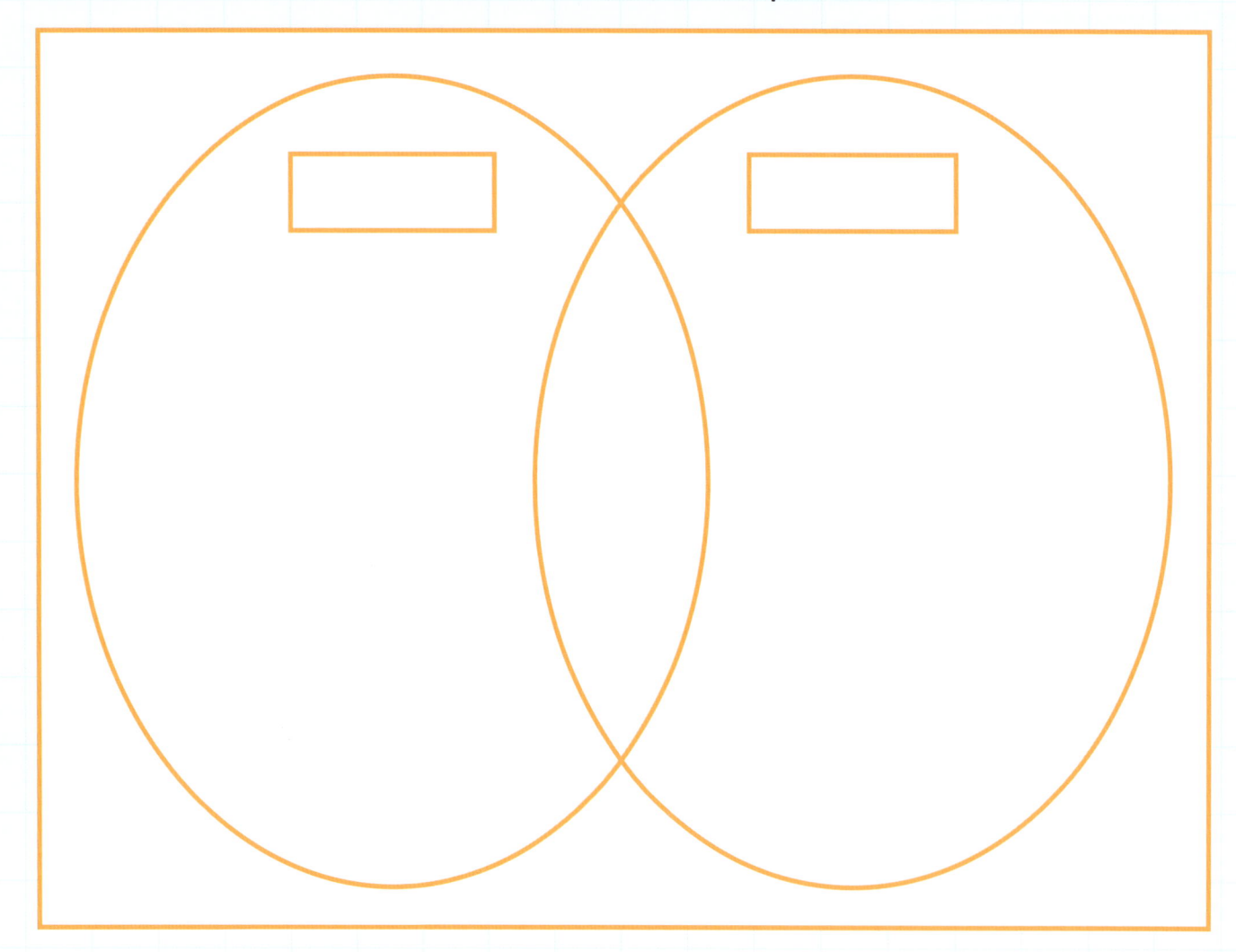

Complete these sentences.

We are both _____.

We both like _____.

Neither of us _____.

12C Sorting using Venn diagrams

Explore

First, label the whole diagram with the range of numbers you want to think about.

Next, label each oval. Your labels could be odd, even, numbers over 20, multiples of 5 or something else.
Now sort your numbers.

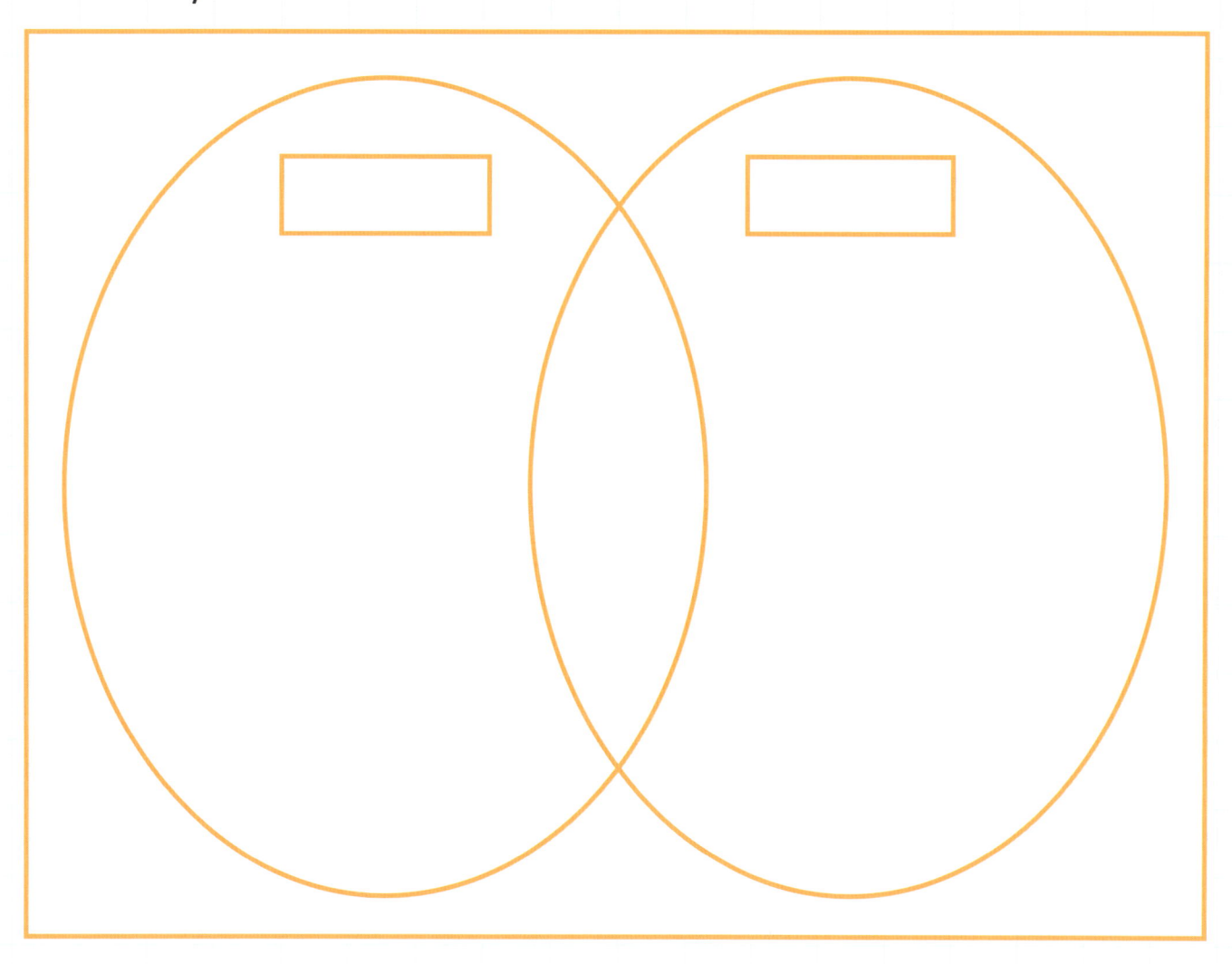

Numbers in the overlap are _____ and _____.

12 Handling data

Connect

Carry out a survey of what students in your class like to do after school. What are the favourite after-school activities?

What else does the information tell you?

Use a Carroll diagram to sort for two criteria. This could be mammals/not mammals with four legs/not four legs; students who have brothers/no brothers with students who have sisters/no sisters; brown eyes/not brown eyes with fair hair/not fair hair or something else you are both interested in.

Use a Venn diagram to show how some information belongs in one oval, some information belongs in the other oval and some information belongs in both. You could explore what you and your partner like, friends who can ride a bike and/or swim, land and sea animals, straight and curved lines in capital letters, something about numbers or shapes or something else you are both interested in.

12 Handling data

Review

Draw a Venn diagram using the information in the Carroll diagram below. Remember to label the ovals and the box around the Venn diagram.

	Multiple of 2	Not a multiple of 2
Multiple of 5	10 50 80	45 25 5
Not a multiple of 5	42 36 28	33 27 51

All the students in Class 2 were asked their favourite colours. The teacher collected the information in a tally chart. Draw the tally chart to match the block graph.

Favourite Colour

Red	7
Blue	4
Green	3
Yellow	1
Orange	2
Purple	6

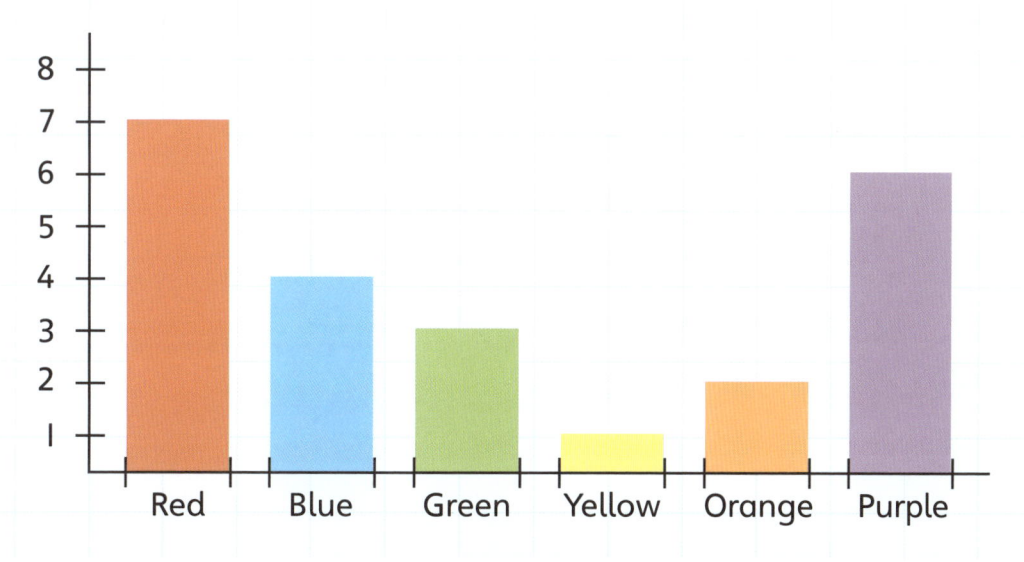

Now draw a pictogram showing the same information. What simple picture or shape will you use in your pictogram?

How many students were asked about their favourite colour?

What are the two most popular colours? _____

Glossary

addition

$4 + 6 = 10$

This is an **addition**

analogue clock

an **analogue** clock   an **analogue** watch

anticlockwise/clockwise

clockwise   anticlockwise

array

an **array**

block graph

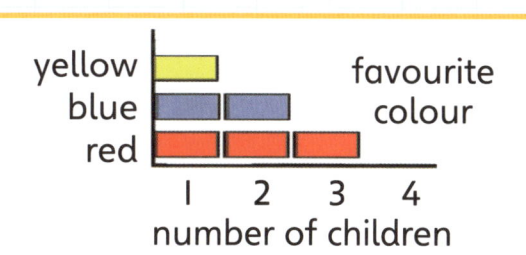

favourite colour

number of children

This **block graph** tells you how many children like each colour

calculate

Can you **calculate** the answers?

$4 + 3$
$8 - 4$
2×3

$4 + 3 = 7$ $8 - 4 = 4$ $2 \times 3 = 6$

187

capacity

This bucket has a **capacity** of 8 litres

centimetre/metre

circular

These shapes have **circular** tops

column

1	2	3	4
5	6	7	8
9	10	11	12
13	14	15	16

a **column**

digital clock

a **digital clock**

a **digital watch**

divide

You can **divide** 6 into two equal groups of 3

188

exactly

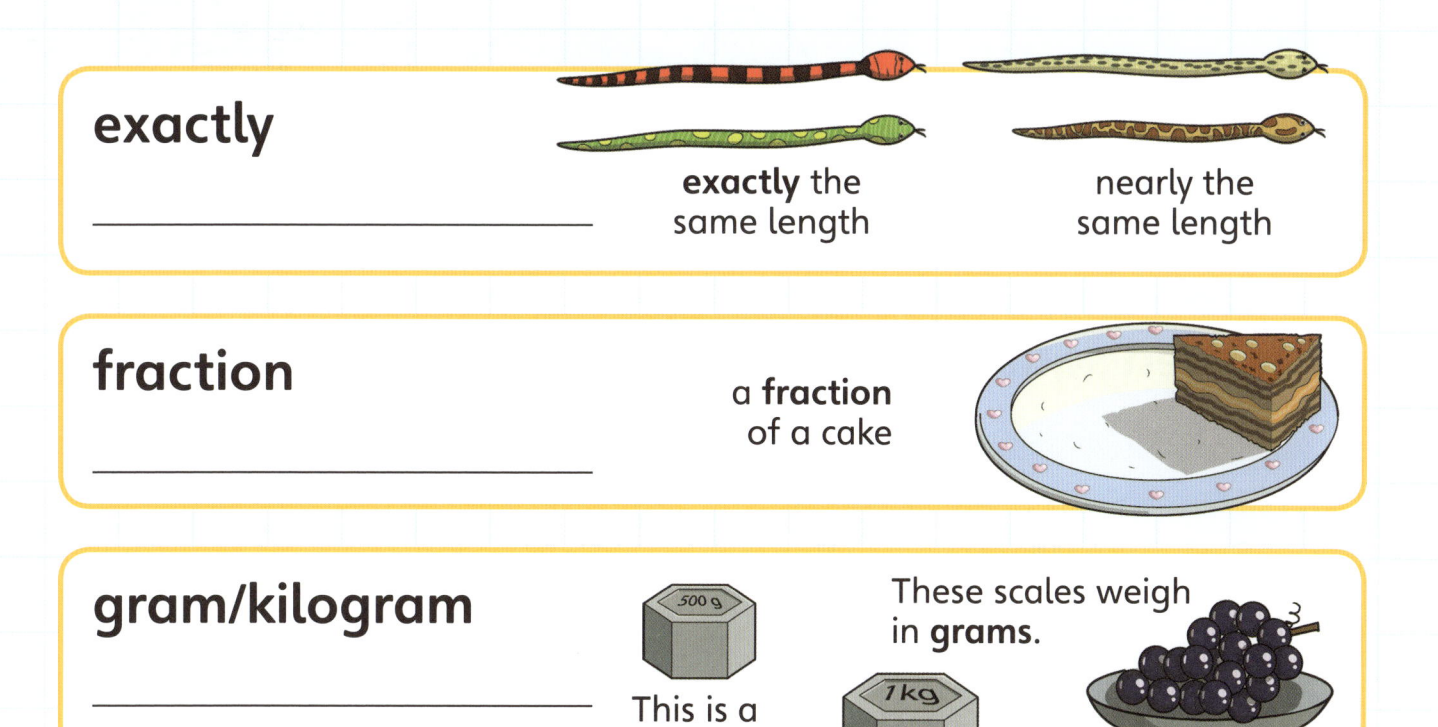

exactly the same length

nearly the same length

fraction

a **fraction** of a cake

gram/kilogram

This is a 500 **gram** weight.

a 1 **kilogram** weight

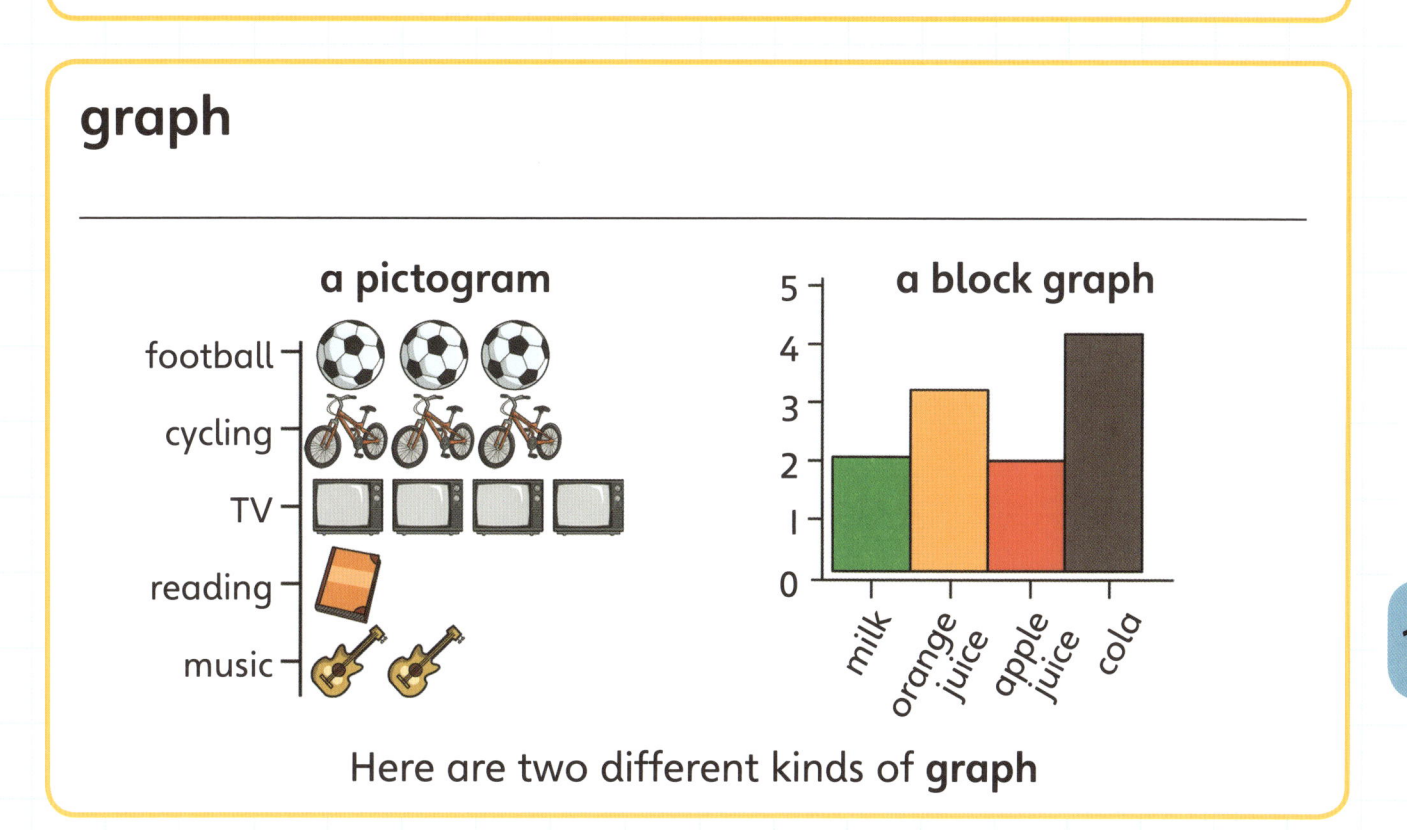

These scales weigh in **grams**.

graph

a pictogram

football
cycling
TV
reading
music

a block graph

milk
orange juice
apple juice
cola

Here are two different kinds of **graph**

Glossary

189

hexagon

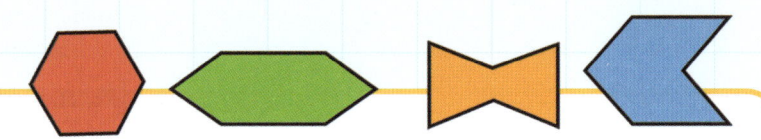

All of these are **hexagons**. The red shape is a regular **hexagon**

hundred square

1	2	3	4	5	6	7	8	9	10
11	12	13	14	15	16	17	18	19	20
21	22	23	24	25	26	27	28	29	30
31	32	33	34	35	36	37	38	39	40
41	42	43	44	45	46	47	48	49	50
51	52	53	54	55	56	57	58	59	60
61	62	63	64	65	66	67	68	69	70
71	72	73	74	75	76	77	78	79	80
81	82	83	84	85	86	87	88	89	90
91	92	93	94	95	96	97	98	99	100

line of symmetry/ mirror line

a line of symmetry

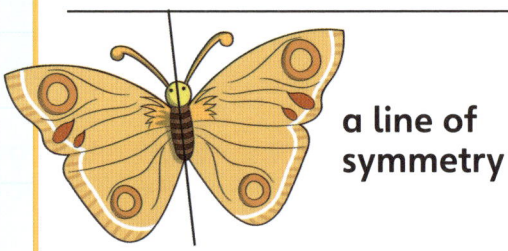

A line of symmetry is like a mirror because one side is a reflection of the other

investigate

least

the **least** juice

litre/millilitre

one **litre** of milk

5 **millilitres**

minute
_____
_____

Five **minutes** have passed.

most
_____
_____

Farrah   Sara   Mustafa

Mustafa has the **most** apples

multiple
_____
_____

multiples of 2	2, 4, 6, 8, 10 …
multiples of 5	5, 10, 15, 20, 25 …
multiples of 10	10, 20, 30, 40, 50 …

The **multiples** go on and on

multiply
_____
_____

Multiply 3 by 4. What is the answer?

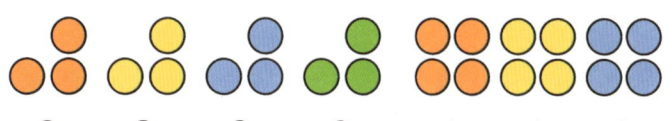

3 + 3 + 3 + 3   or   4 + 4 + 4

The answer is 12.

number bond
_____
_____

addition bond	3 + 4 = 7
subtraction bond	5 − 1 = 4
multiplication bond	2 × 3 = 6
division bond	10 ÷ 2 = 5

191

number grid

1	2	3
4	5	6
7	8	9

Numbers arranged in a **grid**

number pair

5 and 0   3 and 2   4 and 1

All these **number pairs** total 5

octagon

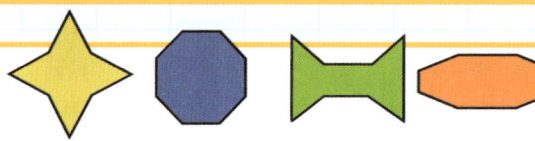

All of these are **octagons**. The blue shape is a regular **octagon**

one-digit number, two-digit number, three-digit number

7 — a one-digit number
36 — a two-digit number
248 — a three-digit number

part/whole

4 equal **parts**

1 **whole**

whole   **halves**

quarters

pentagon

All of these are **pentagons**. The blue shape is a regular **pentagon**

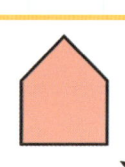

192

place value

1327

value 1000   value 300   value 20   value 7

predict

I **predict** that the next bead is yellow, because the pattern is yellow, blue, yellow, blue.

quarter

quarters   not **quarters**

rectangular

The football field is **rectangular**

right angle

a whole turn   a quarter turn   a **right angle**

193

round

22 rounded to the nearest ten is 20.

389 rounded to the nearest hundred is 400.

row

a row

1	2	3	4
5	6	7	8
9	10	11	12
13	14	15	16

Which are the even numbers in this **row**?

The answer is 6 and 8.

rule

The **rule** for this machine is add 2

4 +2 6

sequence

3 6 9 12 15 →

second

second hand

share

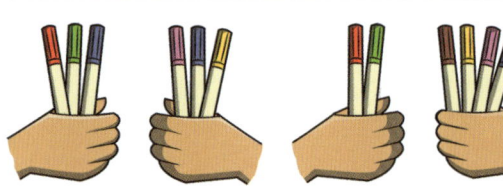

6 **shared** equally

6 **shared** unequally

solve

Solve this problem: I have 7 sweets. If I eat 3 how many will I have left?

7 − 3 = 4. You will have 4 left.

straight line

The shortest line joining two points

subtraction

7 − 4 = 3

This is a **subtraction**

surface

top
the outer area
bottom

A cylinder has two flat **surfaces** and one curved **surface**

symbol

Here are some **symbols**. Do you know what they all mean?

×   cm   kg   ÷

× multiply   cm centimetre   kg kilogram   ÷ divide

195

Glossary

symmetrical
_____
_____

The ladybird is **symmetrical**

tally
_____
_____

The **tallies** show that 4 people ate sandwiches

triangular
_____

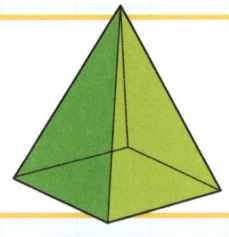

turn
_____
_____

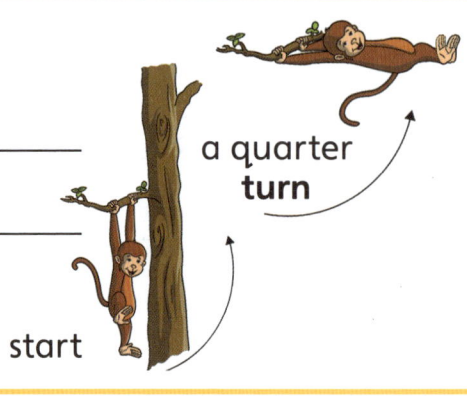

start

a quarter **turn**

a half **turn**

a whole **turn**

whole number
_____

2 is a **whole number**

$\frac{1}{2}$ is a fraction

$1\frac{1}{2}$ is a mixed number.

196